Understanding Mathematics and Science Matters

Understanding Mathematics and Science Matters

Edited by

Thomas A. Romberg
Thomas P. Carpenter
Fae Dremock
University of Wisconsin–Madison

Routledge
Taylor & Francis Group
New York   London

The research reported here was supported in part by a grant from the National Science Foundation (ESI9911679) and a grant from the Department of Education Office of Educational Research and Improvement to the National Center for Improving Student Learning and Achievement in Mathematics and Science (R305A60007-98). The opinions expressed in this book do not necessarily reflect the position, policy, or endorsement of the supporting agencies.

First published by Lawrence Erlbaum Associates, Inc., Publishers
10 Industrial Avenue
Mahwah, New Jersey 07430

Transferred to digital printing 2010 by Routledge

Routledge

270 Madison Avenue
New York, NY 10016

2 Park Square, Milton Park
Abingdon, Oxon OX14 4RN, UK

Cover design by Kathryn Houghtaling Lacey

Library of Congress Cataloging-in-Publication Data

Understanding mathematics and science matters / Thomas A. Romberg, Thomas P. Carpenter, and Fae Dremock, Eds.
    p. cm. — (Studies in mathematical thinking and learning)
    Includes bibliographical references and index.
ISBN 0-8058-4694-8 (cloth : alk. paper)
ISBN 0-8058-4695-6 (pbk. : alk. paper)
1. Mathematics—Study and teaching. 2. Science—Study and teaching. I. Romberg, Thomas A. II. Carpenter, Thomas P. III. Dremock, Fae. IV. Series.
QA11.2.U53 2005
510'.71—dc22
                        2004053283
                        CIP

10 9 8 7 6 5 4 3 2 1

Contents

Preface

The research reported in this book focuses on providing reliable evidence on and knowledge about forms of mathematics and science instruction that focus on student understanding. The chapters in the book respond to the need to shift the conception of mathematics and science instruction in schools from the "assembly-line" metaphor consistent with the mechanistic views of the 19th century to a conception of mathematical and scientific literacy based on an "exploring a domain" metaphor consistent with the dynamic views of the emerging technological world. To provide a research basis for this shift in instructional practices, the National Center for Improving Student Learning and Achievement in Mathematics and Science (NCISLA), established in 1996 as a research center of the U.S. Department of Education, worked to build the foundation for such reform in the United States based on the recommendations made during the past decade by the American Association for the Advancement of Science (1989), the National Council of Teachers of Mathematics (1989, 1991, 1995, 2000), and the National Research Council (1996).

Developing the conceptions of classroom instruction involves instructional design as well as examination of student learning and teachers' practices, all of which are concerns of professional development. In fact, to conduct research it was not sufficient to study instruction as it currently exists in schools; rather, we studied classrooms in which compelling new visions of mathematics and science are becoming the norm. In order to enable other classrooms to implement these visions, we were seeking to understand not only how these classrooms function, but also what it takes to construct and maintain such classrooms. Creating such classrooms has involved the coordinated work of teachers and researchers *working together*. The purpose of this book is to present a summary of the concepts, findings, and conclusions of our research and, in so doing, provide others with a concrete picture of what the mathematics or science classroom of the future could be like.

Our research focuses on the empirical development of conceptual frameworks ("causal models" or "local instructional theories"). These studies differ from the idealized top-down psychological teaching experiment, which carries out the design step in a laboratory setting and then attempts to transfer the design to "the 'messy' context of an ongoing classroom" (Klahr, Chen, & Toth, 2001, p. 85). Of course, the fabrication of a sequence of instructional activities is constrained by such practical considerations as grade level, assumed prior knowledge, and so on, as well as the focus on student understanding in the domain as an outcome. However, classroom experimentation is not carried out simply to demonstrate the effectiveness of the designed activities. Instead, a defining characteristic of formal design experiments is the ongoing cycle of design and analysis (Gravemeijer, 1998). Cobb (2001) argued that an instructional experiment should "result in analyses that feed back to inform the improvement of instructional designs," and "enable documentation of the developing mathematical [and scientific] reasoning of individual students as they participate in the practices of the classroom community" and of "the collective mathematical [and scientific] learning of the classroom community over the extended periods of time spanned by design experiments" (p. 463). When conducting such experiments, research teams spend time in classrooms attempting to understand and to improve the collective process of teaching and learning mathematics or science in specific domains. The outcome of such work includes sequences of tasks, tools, activity structures; norms that support such teaching and learning; and analyses of the process of student learning as a consequence of classroom instruction using those activities and tools.

Teachers play a critical role in these studies because they foster the collaboration of teachers and researchers. In particular, the teachers' role in creating the general classroom participation structure, negotiating the use and meanings of terms and strategies, and implementing formative evaluation procedures is critical. Given that this role is one they generally are not accustomed to, intensive professional development has been commonly integrated into the studies reported in this book.

Teaching for understanding and implementing model-based reasoning and argumentation involves a significant reorientation of teacher beliefs and the acquisition of new forms of pedagogical and content knowledge. Traditional efforts to help teachers develop as professionals, such as one-shot workshops or simple declarations about the need to reform, are inadequate and contradict what we understand about human learning. Our work has focused on various means to help teachers establish close and clear relationships between their ongoing work in classrooms and their efforts to reform mathematics and science instruction.

The work reported in this book goes far beyond mere opinion, guesswork, or flight of fancy. Granted, findings do not typically stay "found" or "complete": The evidential base—what constitutes a reasonable argument and the given purposes—continually changes. Just as today's students are far removed from those in classrooms in the 1950s in culture and mores, what they are learning—and need to learn in order to be productive citizens—has also changed. Reaching all students and supporting them intellectually as they reach for technologically demanding and very different standards of achievement and knowledge requires more-inclusive teaching in the classroom and more individualized and interactive testing.

Although many schools and teachers are convinced of the need for reform in the teaching and learning of mathematics and science, they are struggling with conflicting demands, lack of relevant materials, limited staff development resources, and personal concerns about the consequences of taking risks. What they need is reliable information about what can be accomplished, realistic examples of classrooms in which instruction promotes understanding, convincing evidence that such instruction yields high achievement, and clear recommendations of what they can do based on a coherent set of design principles. As a consequence of our work, we are confident that practitioners, policymakers, and the public can be helped to see the validity of the reform recommendations, understand recommended guidelines, and use these to transform the teaching and learning of mathematics and science in U.S. classrooms.

OVERVIEW OF THE CHAPTERS

Understanding Mathematics and Science Matters is divided into four parts:

- *Part I: Introduction* contains a single overview chapter that situates the chapters in this book in terms of the reform movement in school mathematics and school science.
- *Part II: Learning With Understanding* contains seven chapters that focus specifically on research directed toward what is involved when students learn mathematics and science with understanding.
- *Part III: Teaching for Understanding* contains three chapters that focus the research on the role and problems teachers face when attempting to teach their students mathematics and science with understanding.
- *Part IV: Cross-Cutting Studies* includes two chapters that involved gathering information in collaboration with the authors of several other

chapters about classroom assessment practices and organizational support for reform.

—*Thomas A. Romberg*
—*Thomas P. Carpenter*
—*Fae Dremock*

REFERENCES

American Association for the Advancement in Science. (1989). *Science for all Americans*. Washington, DC: Author.

Cobb, P. (2001). Supporting the improvement of learning and teaching in social and institutional context. In S. Carver & D. Klahr (Eds.), *Cognition and instruction: Twenty-five years of progress*. Mahwah, NJ: Lawrence Erlbaum Associates.

Gravemeijer, K. (1998, April). *Developmental research: Fostering a dialectic relation between theory and practice*. Paper presented at the research presession of the annual meeting of the National Council of Teachers of Mathematics, Washington, DC.

Klahr, D., Chen, Z., & Toth, E. (2001). Cognitive development and science education: Ships that pass in the night or beacons of mutual illumination? In S. Carver & D. Klahr (Eds.), *Cognition and instruction: Twenty-five years of progress* (pp. 75–119). Mahwah, NJ: Lawrence Erlbaum Associates.

National Council of Teachers of Mathematics. (1989). *Curriculum and evaluation standards for school mathematics*. Reston, VA: Author.

National Council of Teachers of Mathematics. (1991). *Professional standards for teaching mathematics*. Reston, VA: Author.

National Council of Teachers of Mathematics. (1995). *Assessment standards for school mathematics*. Reston, VA: Author.

National Council of Teachers of Mathematics. (2000). *Principles and standards for school mathematics*. Reston, VA: Author.

National Research Council. (1996). *National science education standards*. Washington, DC: National Academy Press.

Part I

INTRODUCTION

Chapter 1

Standards-Based Reform and Teaching for Understanding

Thomas A. Romberg
Thomas P. Carpenter
Joan Kwako
University of Wisconsin–Madison

The U.S. "standards-based" reform movement in school mathematics and science has been based on the belief that, in most classrooms at all school levels, mathematics and science instruction is neither suitable nor sufficient to adequately equip our children with the concepts and skills needed for the 21st century. Furthermore, unless something is done to alter current schooling trends, conditions are likely to get worse in the coming decades. In 1983, the need for reform in classroom practices was brought vividly to the attention of the U.S. public with the publication of *A Nation at Risk* (National Commission on Excellence in Education, 1983) and *Educating Americans for the 21st Century* (National Science Board Commission on Precollege Education in Mathematics, Science, and Technology, 1983). The authors of those documents claimed that competing in a global environment depended on a workforce knowledgeable about the mathematical, scientific, and technological aspects of the emerging information age and that our schools were failing to prepare students for their future. This concern was reechoed in *Before It's Too Late: A Report to the Nation* (National Commission on Mathematics and Science Teaching for the 21st Century, 2000) and in the No Child Left Behind Act of 2001.

The failure to educate our students was seen as a product of the traditional, if simplistic, view of learning and teaching commonly practiced in

U.S. schools. Both school mathematics and science were seen as a long list of established concepts and procedures. Learning was seen as acquiring these concepts and procedures through memorization and repeated practice. The job of teaching involved planning, presenting, and keeping order; a teacher's responsibility basically ended when he or she told students what they had to remember. The school mathematics and science curricula were characterized as superficial, underachieving, and diffuse in content (McKnight & Schmidt, 1998; Peak, 1996). As a consequence, school mathematics and science in most classrooms was a tedious, uninteresting path to follow, with numerous hurdles to clear. It bore little resemblance to what a mathematician, a scientist, or user of either discipline does.

The goal of the research at the National Center for Improving Student Learning and Achievement in Mathematics and Science (NCISLA) was to study and document the effects of alternative notions of the ways school mathematics and science can be organized and taught so that students learn, with understanding, the concepts, skills, and practices of science and mathematics.

MATHEMATICAL AND SCIENTIFIC LITERACY

Our vision of alternative instruction in mathematics and science is based on the expectation that students should become mathematically and scientifically literate. Language literacy, or the ability to read, write, listen, and speak a language is the primary tool through which human social activity is mediated. Each human language and each human use of language has both an intricate *design* and a variety of *functions*, which are linked in complex ways (Gee, 1998). For a person to be literate in a language implies that the person knows many of the design resources of the language and is able to use those resources for several different social functions. Analogously, considering mathematics or science as a language implies that students must learn not only the concepts and procedures of specific mathematics or science domains (its design features), but also how to use such ideas to solve nonroutine problems and to model or mathematize in a variety of situations (its social functions). For example, when learning a foreign language one must learn the nouns, verbs, and unique sentence structure (its design features), and one must learn how these are used to communicate in a variety of different social situations (e.g., carry on a discussion, ask questions, read a newspaper). Similarly, when learning algebra one must learn how to write and graph linear equations (design features) and to use such equations to represent a variety of contextual problems involving the joint variation of two variables. Developing the languages of mathematics and science and learning to use them in a variety of situations are essential ways of understanding the worlds we experience. Several chapters in this book address issues associated with student appropriation of the languages of mathematics and science. For example, in chapter 3 Rosebery, Warren, Ballenger, and Ogonowski examine the relationship between "everyday" and "scientific" knowledge and learning, and in chapter 8 Nemirovsky, Barros, Noble, Schnepp, and Solomon argue the idea that learning mathematics entails becoming familiar with "symbolic places."

The National Council of Teachers of Mathematics (NCTM) used this analogy in 1986 when it charged the Commission on Standards for School Mathematics to do the following:

> • Create a coherent vision of what it means to be mathematically literate both in a world that relies on calculators and computers to carry out mathematical procedures and in a world where mathematics is rapidly growing and is extensively being applied in diverse fields; and
> • Create a set of standards to guide the revision of the school mathematics curriculum and its associated evaluation toward this vision. (NCTM, 1989, p. 1)

This commission (chaired by Thomas Romberg and contributed to by several NCISLA researchers) produced NCTM's (1989, 1991, 1995, 2000) four standards documents.

Similarly in science, both the American Association for the Advancement of Science (1993) and the National Research Council (1996; report produced by a group chaired by former NCISLA Associate Director Angelo Collins) stressed science literacy as the primary goal for reform in science teaching. The Organization for Economic Cooperation and Development (OECD) also used this analogy in its development of instruments to monitor reading literacy, mathematical literacy, and scientific literacy for its Program for International Student Assessment (PISA; OECD, 1999, 2001, 2002). (NCISLA PI Jan de Lange chaired the Mathematics Functional Expert Group, of which Thomas Romberg was a member, that produced the mathematics framework and test items for this study.)

To become literate in mathematics and science requires a shift in epistemology about the learning of mathematics and science and, as a consequence, a shift in pedagogy. The epistemological shift involves moving from judging student learning in terms of mastery of concepts and procedures to judging student understanding of the concepts and procedures and student ability to scientifically model or to mathematize problem situations. For example, in many of the chapters in this book, instruction involves posing open contextual problems for students (a social function) in order to generate a need for ways to communicate their notions (develop design features).

FEATURES OF MATHEMATICAL AND SCIENTIFIC LITERACY

In this section, we explore the idea of scientific and mathematical literacy by considering how instruction might address the content and practices of these disciplines.

Domain-Based Knowledge

To teach for mathematics and science literacy, the first issue to be addressed involves deciding what content students should understand. Our strategy has been to define the range of content in a domain by using a phenomeno-logical approach to describing the mathematical and scientific concepts, structures, or ideas in relation to domain phenomena and the kinds of problems commonly used. Note that although both mathematics and science have long been described in terms of common general topics (e.g., arithmetic, algebra, geometry; biology, chemistry, physics), the domain approach focuses on problem areas (big ideas) that give rise to those topics. This conception is based on the fact that mathematics and science are composed of diverse domains of knowledge in response to particular problem areas. Much of the Center's work focused on the important features of specific content domains in mathematics and science that we expect students to learn with understanding (e.g., univariate and bivariate exploratory data analysis in chap. 6, arithmetic equivalence in chaps. 4 and 5, genetics and EMS astronomy in chap. 7).

We expect students to gradually acquire the concepts and skills in a domain as a consequence of solving problems, be able to relate those ideas to each other (and other ideas in other domains), and to use those ideas in new problem situations. This phenomenological organization for mathematical content is not new. Two well-known publications, *On the Shoulders of Giants: New Approaches to Numeracy* (Steen, 1990) and *Mathematics: The Science of Patterns* (Devlin, 1994), described mathematics in this manner. As Hans Freudenthal (1983) observed, "Our mathematical concepts, structures, ideas have been invented as tools to organize the phenomena of the physical, social, and mental world" (p. ix). Kitcher (1993) described science similarly in *The Advancement of Science: Science Without Legend, Objectivity Without Illusions*. In keeping with this work, we identified in each domain the key features and resources important for students to discover, use, or even invent for themselves.

This domain view of mathematics and science differs from current perspectives in at least three important ways. Initially, the emphasis is not with the parts of which things are made, but with the whole of which they are part (i.e., how concepts and skills in a domain are related) and in turn how those parts are related to other parts, other domains, and ideas in other disciplines. Secondly, this conception rests on the signs, symbols, terms, and rules for use, that is, the language that humans have invented to communicate with each other about the ideas in the domain. Students should experience the need for the elements of the languages of mathematics and science; as a consequence, teachers must introduce and negotiate with the students the meanings and use of those elements. In chapter 3, Rosebery et al. illustrates the embodied imagining students use in attempting to understand and explain motion down a ramp. In this example, as well as many others, the choice of scientific and mathematical methods and representations is often dependent on the situations in which the problems are presented.

Pedagogically, the domain view of mathematics and science requires a shift from the "assembly-line" metaphor based on assumptions prevalent in the industrial age of the past century to a "domain-based" metaphor. Greeno (1991) argued that students develop understanding when

> a domain is thought of as an environment, with resources at various places in the domain. In this metaphor, knowing is knowing your way around in the environment and knowing how to use its resources. This includes knowing what resources are available in the environment as well as being able to find and use those resources for understanding and reasoning. Knowing includes interactions with the environment in its own terms—exploring the territory, appreciating its scenery, and understanding how its various components interact. Knowing the domain also includes knowing what resources are in the environment that can be used to support your individual and social activities and the ability to recognize, find, and use those resources productively. Learning the domain, in this view, is analogous to learning to live in an environment: learning your way around, learning what resources are available, and learning how to use those resources in conducting your activities productively and enjoyably. (p. 175)

In particular, in chapter 8, Nemirovsky et al. use the metaphor of becoming familiar with a new place (e.g., a town, a train station) and the process of learning a new mathematical concept.

Learning Corridors

In planning instruction, curriculum developers often refer to scope and sequence, suggesting that both what is taught and the order in which it is taught are important. Center researchers have addressed the issue of sequence of instruction that leads to student understanding of science and mathematics. Our conception is based on the notion that an instructional sequence for students and teachers should follow a variety of paths along a "broad corridor" (Brown 1992; Brown & Campione, 1996). The notion of "learning corridor" has also been called "hypothetical learning trajectory" (Cobb, 2001; Simon, 1995), both to emphasize the speculative nature of the possible paths and to focus on the trajectory because the "learning for understanding" perspective sees students progressing from informal ideas in a domain to more formal ideas over time and, in so doing, reconfiguring their knowledge by developing strategies and models that help them discern when and how facts and skills are important. For example, in chapter 6, McClain, Cobb, and Gravemeijer describe the task sequence that evolved in their study of students' learning of statistical data analysis.

To engineer classroom instruction based on the domain-based metaphor, a collection of problem situations is needed that engage students in exploring the environment of each domain in a structured manner. Students then have the opportunity to construct their mathematical or scientific knowledge from these purposeful activities. To accomplish this, two distinct dimensions need to be considered. The first is that the domain needs to be well mapped. This is not an easy task, as Webb and Romberg (1992) argued:

> [K]nowledge of a domain is viewed as forming a network of multiple possible paths and not partitioned into discrete segments.... Over time, the maturation of a student's functioning within a conceptual field should be observed by noting the formation of new linkages, the variation in the situations the person is able to work with, the degree of abstraction that is applied, and the level of reasoning applied. (p. 47)

The second is that activities that encourage students to explore the domain need to be identified and organized in a structured manner that allows for such learning. Doing both is not easy. For example, although there is no doubt that many interesting activities exist or can be created, whether they lead anywhere is a serious question. Keitel (1987) argued that an activity approach in mathematics can lead to "no mathematics at all," and Romberg (1992) noted that

> too often a problem is judged to be relevant through the eyes of adults, not children. Also, this perception is undoubtedly a Western, middle-class, static vision. Concrete situations, by themselves, do not guarantee that students will see relevance to their worlds; they may not be relevant for *all* students nor prepare them to deal with a changing, dynamic world. (p. 778)

New instructional materials should provide opportunities for students to be active and should constantly extend the structure of the mathematics and science that students know by having them make, test, and validate models and conjectures within and across domains. Clearly, the work of students should no longer be a matter of acting within somebody else's structures, answering somebody else's questions, and waiting for a teacher to check the response.

Growth in a domain depends on learning to participate in a broad range of social situations by recognizing the similarities and differences between practices in various situations and being able to use aspects of practices learned in one situation for successful participation in other situations. Because learning in a domain occurs over several years, instruction should also consider the learning histories of students and arrange conditions for students to progressively build and cumulate knowledge. In this sense, understanding in a domain, as Shafer and Romberg (1999) stated,

> develops as new relationships among pieces of existing knowledge are sought, tested, and realized, or as new information is connected to and integrated with existing knowledge. As understanding grows, the facts, relationships, and procedures in a domain become resources that aid reasoning in solving ordinary problems in routine ways and in generating insights for making sense of unfamiliar situations. This growth in understanding in a domain occurs intermittently and sporadically in periods of progression and regression, rather than in linear increments or stages. (pp. 159–160)

Thus, movement along a learning corridor is not necessarily smooth, or all in one direction. Students move back and forth from informal to formal notions depending on the problem situation or on the mathematics or science involved.

Scientific and Mathematical Practice

This conception of what to consider when developing instruction that leads to understanding elaborates on the "social function" idea about mathematical and scientific literacy by focusing on what the work of scientists and mathematicians actually entails. Most young children perceive science largely as a passive process of observing and recording events, and mathematics as a process of learning calculation procedures for bigger and bigger numbers (Songer & Linn, 1991). High school students often believe that good scientists are those who attend acutely and complete records of all they observe (Carey, Evans, Honda, Jay, & Unger, 1989) and that good mathematicians are those who can carry out complex routines accurately.

Yet accounts of the work of professional scientists and mathematicians present quite a different picture, one dominated by building and testing models, by mathematizing nonroutine problems (Hestenes, 1992; OECD, 2002), and, as Stewart, Cartier, and Passmore (2001) noted, by "joining causal models with empirical data in order to construct explanations" (p. 3). As Stewart et al. (2001) also noted, by embracing the notion of scientific and mathematical practice, "curriculum developers and teachers ... can create instruction that permits students to go beyond learning a formulaic set of rules to be unthinkingly followed, and beyond the rather oversimplified 'understanding' of science [and mathematics] that results" (p. 6). Students are, as Freudenthal (1983) noted, "entitled to recapitulate in a fashion the learning process of mankind" (p. ix), and the activities that they engage in can more effectively arise "from reality itself, which is not a fixed datum, but expands continuously in individual and collective learning process" (Freudenthal, 1987, p. 280). The mathematician George Polya (attributed by Kilpatrick, 1987) felt that we understand mathematics (and science) best when we see it being born, by either following in the steps of historical discoveries or by engaging in discoveries ourselves. From this perspective, instructional activities should allow students to become re-inventors of mathematical and scientific concepts, with teachers guiding the process and making students conscious of the terms, signs, symbols, and rules for use so that they can reflect on the process.

Although relatively few students go on to become scientists or mathematicians, it is our belief that in our increasingly complex world all students should experience the practice of scientific and mathematical inquiry. To fully understand mathematics and science, and the policy issues in these areas, students must participate in the practices used by people who engage in generating and using mathematical and scientific knowledge. In the past, these practices have often been implicit in student activity, but because they were not made an explicit focus of attention, they were often unnoticed and not learned by most students.

MODELS AND MODELING

Meaningful inquiry, involving cycles of model construction, model evaluation, and model revision, is central to understanding in a domain and fundamental to the professional practice of mathematicians and scientists. The use and diversity of models in these disciplines suggest that modeling can help students develop understanding of a wide range of important mathematical and scientific ideas, and that modeling practices can and should be fostered at every age and grade. Our research has convinced us that instruction that focuses on students modeling phenomena is a powerful way to accomplish learning with understanding in mathematics and science classrooms.

Perspectives on Models and Modeling

Although different conceptions or types of models can be distinguished in a variety of ways, for our purposes we distinguish between "model as *a natural process* used to construct an explanation of natural phenomena" and "model as *a representational tool* for communicating about the conceptual referent." Both conceptions of models begin with a phenomenological context (e.g., event, question, problem situation), both involve identifying key features or attributes of the phenomena and how those features are related, and both use representations as tools to support disciplinary practices such as communication, mobility, combination, selection, prediction, and so on. However, these concepts differ in the use of the term *model*, the focus of emphasis, and the ways models are validated.

Models and Modeling as Process. In science, a "model is a set of ideas that describes a natural process" (Cartier, Rudolph, & Stewart, 2001, p. 2). In this sense, a model is a "theoretical world" from which explanations are constructed. Such conceptual models are often characterized as "scientific models," "explanatory models," "causal models," or "models for." Conceptual models can be created for anything, ranging from the economics of moving icebergs (Cross & Moscardini, 1985) to the role of intuition in the process of creating knowledge (Kuyk, 1982). Note that in this conception, although "representations of models [e.g., formulae, physical replicas] are essential tools for communicating and conversing about the conceptual models ... we take the position that representations are not models themselves" (Cartier et al., 2001, p. 2). From this perspective, the practice of modeling involves the process of development and use of a conceptual model and includes problem identification, gestation through reflection, model "building," simulation, and payoff.

The work of scientists and mathematicians involves cycles of model development, evaluation, and revision and arises out of phenomena for which an "explanation" is sought. Initially, because phenomenological situations are rarely well defined, this work means the formulation or identification of a question to be "answered" or a data pattern to be explained. As a part of the process of identifying the problem, one attempts to sort out the set of key concepts or significant features of the situation. This simplification or idealization is a crucial stage because often the general problem is exceedingly complex. In particular, it involves making decisions about what can be ignored and developing a sense about how the essential concepts or features are connected. Once the set of ideas has been identified, the next stage is to describe the set of ideas via some representation to communicate with others the features of the model. The development of representations is critical because it can often facilitate inquiry into and understanding of the original problem situation.

Model evaluation, which follows, usually involves generating hypotheses from which empirical studies follow and can also involve manipulating or reasoning with representations (sometimes expressed mathematically) to deduce conclusions or conjectures and to examine the validity of assertions. Some form of validation is usually carried out throughout the formulation (i.e., the formulae or other mathematical relations set up to represent the model are continually checked with the initial situation), and the mathematics used in the model must obey all the usual laws of mathematical logic. A model's validity rests, however, in its ability to adequately explain the data gathered from the problem situation initially described.

In summary, from this perspective, models are conceptual systems that represent phenomena in the world by means of a system of theoretically specified objects, relations, operations, and rules governing interactions. Models are not simply physical images; they are ideas. Models may be represented by physical or iconic images, or mathematical symbols, but these representations are not models themselves; rather, they represent ideas at the heart of the model.

Model and Modeling as Representations and Inscriptions.   Lehrer and Schauble (2003), on the other hand, argued that models *are* the representations or inscriptions constructed (or adopted) as conventions within a community to support disciplinary practices. Such physical models are often characterized as "descriptive models" or "models of." The process includes:

- The explicit and conscious *separation* of a model and its referent.
- The comparison of two or more models for their *relative fit* to the world.
- The role of *mathematization* in modeling.
- The experience in *conventionalizing* representations, that is, using notations and inscriptions that are adopted as conventions for supporting the reasoning of a practicing community.

In this perspective, modeling should be viewed as a complex process involving the creation of representations (sometimes using the specialized language of mathematics) to convey information about the important features of a problem situation. Although creating a model starts simply, it often develops into a sophisticated tool as greater understanding of the problem situation is achieved. In this sense, any judgment about the validity of a model is not whether it is a "true" model (i.e., accurately representing the working of the system) or not. The judgment is only about the "adequacy" of the model (i.e., whether the results obtained were sufficiently representative of the situation for the purpose of the problem at hand). Thus, a solution to the original problem depends on criteria assessed by the modeler as much as on the features of the initial situation. Simple, adequate, but incomplete representational models are sometimes more useful than elaborate models requiring elegant conceptual justification or complex mathematical analytic procedures for solutions, and are often less costly.

In summary, from this perspective, models are the representations or inscriptions constructed to address a problem or answer a question and embody only the specific features of the situation necessary to solve the problem or answer the question.

Modeling in Classrooms

Although the differences in the "model as process" and "model as representation" are important to understand, they both should be reflected in reform classrooms. Many people believe that doing school mathematics and science involves simply finding answers to set tasks. We note, however, that in both perspectives of models and modeling, the task is not complete simply because an "answer" has been obtained. Initial models are revised, retested, and so forth. Problem solving often "involve[s] several 'modeling cycles' in which descriptions, explanations, and predictions are gradually refined and elaborated" (Lesh & Doerr, 2003, p. 31). Once a model has been satisfactorily refined (i.e., it sufficiently explains, or represents, the phenomenon studied), however, students still need to interpret the results of the model, and communicate and justify their answers.

Using Explanatory Models in the Classroom.   Explanatory models provide students an entrée to the work of scientists. For example, in one project MUSE (see chap. 7) high school biology class, students tackle problems from classical genetics. Using a computer simulation program that generates crossing fruit flies with different characteristics, students work in small groups to collect data to explain inheritance patterns in wing traits or body color. Students modify their understanding of the Mendelian model of simple dominance to account for a more complex inheritance pattern. They generate, test, and revise their models to explain and predict inheritance patterns involving codominance and multiple alleles for a trait.

In one activity, students investigate fruit fly body color expressed in one of four variations: ebony, green, tan, and pallid. Beginning with a (virtual) vial of fruit flies, students use the computer simulation program to systematically perform a wide range of crosses and record results. Some crosses yield results that can be explained by simple dominance; others yield unexpected results. As they attempt to explain the four body colors seen in the fruit fly populations, students realize that they need to revise their basic model of inheritance. Generally one or more groups, after considerable trial and error, revise the basic two-allele-per-trait model to one with three alleles per trait (with only two alleles inherited in any given cross). They also propose that some alleles are fully dominant whereas others are codominant, which might explain the specific patterns observed. They test their revised model by systematically predicting and testing the results of different crosses. When the patterns of color variation in the data generated by the computer simulation program prove consistent with their new model, and can be explained by it, the group designs a physical representation to use to present their model to the rest of the class. The following involves a group that used a triangle as part of their representation:

Jackie:	We came up with this triangle theory. We decided from the first couple of crosses that we did that ebony was dominant and green was dominant because when we bred ebony and ebony, we got ebony and pallid, and we got it in a 3 to 1 ratio, so we figured that ebony was dominant, just like green was dominant.
Ms. J:	So that actually fit the Mendel model, didn't it?
Jackie:	Right. So we used that in part of ours. But then we got some really weird crosses when we bred ebony and green. We ended up getting all four of them [variations of the trait], and that's what this Punnett square is right here. So we figured out that green was a 2,3 and that the ebony was a 1,2, and when we bred them, we ended up getting all four variations that fit on our model.
Ms. J:	Where did you come up with the 3 [the third allele]? Why did you add a 3?
Jackie:	Because there was no way that we could get four different color variations from just two alleles because you only end up with three different combinations. So even if every single one of those was each a color, you still have one color that is coming from nowhere. So we came up with a, maybe there was three, but you only carried two of them. So you still can end up getting all different four color combinations from having three different alleles.

The students' model adequately explained the variation in a complex pattern of data.

Using Representational Models in the Classroom.   Following the "model as representation" perspective in classrooms involves the development of carefully structured physical tasks in which students confront

the need to conceptualize, test, and refine their ideas about the task. Lehrer and Schauble (2003) illustrated the gradual developmental evolution of children creating such models with examples of tasks that elicit such behaviors. For example, physical microcosms are fruitful places to begin "the modeling game" with children because they rely on physical resemblance for modeling the world. When first graders were asked to use springs, dowels, and other materials to build a model that "works like your elbow," almost all students at first included balls and popsicle sticks to represent hands and fingers (Penner, Giles, Lehrer, & Schauble, 1997). Most children insisted on sticking a small ball onto their model to show "the bump where the elbow should be." Only after constructing physical representations that "looked like" their elbows did the children turn their attention to function. The children who were concerned with representing hands and fingers eliminated these features in their second round of physical models when class attention shifted to making a model that bent through the range (and only through the range) of the human elbow. Lehrer and Schauble (in press) argued, "The purpose of emphasizing physical microcosms with novices or young students is to firmly ground children's understanding in familiar objects and events. This familiarity supports mapping from the representing to the represented world by permitting children to rely on literal similarity to facilitate model-world relations" (p. 11).

Representational systems are usually "grounded" in resemblance between the model and the world, but such representations typically undergo fundamental transformations via inscription. Consider, for example, maps, in which the resemblance of the model to the world relies heavily on convention rather than perception. Reading a typical street map depends on knowledge about conventions (e.g., the compass rose, the meaning of different line weights, the presence of a key) that influence our perceptions and often alter them. Lehrer and Schauble (in press) summarized representational systems as being "initially anchored in resemblance and physical models of the world, which are then stretched to mathematical description. Once stretched, the mathematical descriptions take on a life of their own. They become embodiments for other systems, fulfilling their promise as mobile and extensible modeling tools" (p. 15).

Summary. Both perspectives of models and modeling are reasonable bases for instructional activities in reform mathematics and science classrooms. Both give students an opportunity to experience how scientists and mathematicians develop conjectures about the world and how they justify assertions and a base from which to reason and model "data" they encounter in real life, whether, for example, reasoning from a map or exploring the explanation for shadow length (e.g., closeness of light source to object shadowed). As students learn to see and use models as tools, they adapt both process and representation to reasoning about the world.

ARGUMENTATION

The ability to articulate ideas is a benchmark of understanding. Articulation involves the communication of knowledge, either verbally, in writing, or through such means as pictures, diagrams, or models. Central to developing this ability is the use of mathematical and scientific argumentation to support conjectures. Much research in this volume has focused on the development of classroom discourse during the process of learning. For example, in chapter 4 Carpenter, Levi, Williams Berman, and Pligge describe a progression of forms of argument students use to justify generalizations, and in chapter 2 Lehrer and Schauble describe how design elements are used to support an argument.

Forman (2001) focused on developing a theoretical and research-based framework that could be used to study the discourse of inquiry in mathematics and science classrooms and on employing this framework to assess instructional processes. Her approach, which draws from research in developmental psychology, sociolinguistics, and the sociology of science, as well as from research in education, sees the learning of scientific discourse as an aspect of students' socialization within classroom settings. As Stewart et al. (2001) argued,

> Given the central position of explanations in science, we believe that the constructing of explanations should also be prevalent in science instruction, with opportunities provided for students to engage in argumentation related to the need for and the adequacy of various types of explanations.... [A]rgumentation occurs when students interact with one another, often in ways carefully orchestrated by teachers, around diverse but high-stakes issues associated with the generation and validation of scientific claims. For us, argumentation in science classrooms is simply the public face of the activities of inquiry, modeling, and, particularly, explanation. (pp. 12–13)

The genre of classroom discourse, however, can conflict with the genre of scientific argumentation as well as with the genres already mastered by students in their homes and communities. Given such an array of discursive forms, students can get easily confused, especially if the rules for distinguishing and applying them are implicit. Obviously, teachers have an important role to play in making explicit what has been implicit about the genres students are expected to master in order to appear competent in school (Forman, Larreamendy-Joerns, Stein, & Brown, 1998). For example, although in reform classrooms students are often asked to explain their strategies in solving particular problems, van Reeuwijk and Wijers (2003) showed that students often did not have a way to know when they were finished and no way to validate their answers. Expectations, she noted, must be made explicit.

Prior to school entry, children are exposed to justifications and explanations at home; they also are familiar with resolving interpersonal conflicts with peers (Goodwin & Goodwin, 1987). As they move into the school environment, they are exposed to three basic ways to justify assertions (Romberg, 1983): authority validation, empirical validation, and deductive validation. Authority validation is the process of determining validity by relying on some authority, as when a child, for example, checks his or her answer to a problem by comparing it to an answer book or to the teacher's answer. Empirical validation involves representing a proposition with objects, pictures, or other symbols to assist in determining its validity. (For examples of students empirically justifying the generalization that "zero added with another number equals that other number," see Carpenter & Levi, 2000.) Deductive validation is the process of determining validity by an argument based on agreed-on common notions, definitions, axioms, and rules of logic. Although deductive validation is at the heart of mathematics and science, the transition from relying on authority or empirical examples to using general forms of argument is not easy. In chapter 4, for example, Carpenter et al. describe a class developing norms for justifying assertions and note that the students only gradually began to understand the need to go beyond specific examples. In chapter 2, Lehrer and Schauble provide examples of students making transitions from arguments based on empirical examples to those based on proof.

Because scientific argumentation builds on students' previous experience with disagreements and explanations, classroom norms of behavior conducive to achieving consensus through comparing solutions, challenging explanations, and requiring others to defend their positions, particular norms need to be established (Stewart et al., 2001; Yackel & Cobb, 1996; Yackel, Cobb, & Wood, 1991). Classroom conversations often begin with students using natural language, progressively increasing their vocabulary to include mathematically and scientifically powerful terms and descriptions. Eventually students come to recognize what constitutes a valid mathematical or scientific explanation or solution. As students participate in this type of discourse, they become members of a discourse community, developing mathematical and scientific beliefs, understandings, and values that reflect scientific practice (Cobb, Boufi, McClain, & Whitenack, 1997; Silver & Smith, 1996; Yackel & Cobb, 1996).

The forms of validation of propositions in mathematics and science also vary. Formal and informal scientific arguments (central to scientific practice) often are about the relationship between material and symbolic objects (e.g., models). Establishing the validity of these models using standards of evidence is the most important aim of scientific argumentation (Bazerman, 1988). Stewart et al. (2001) posed the following proficiencies, conducted within a scientific practice, that high school students might be expected to acquire:

1. Developing explanations that (a) link causal models and data, (b) unite diverse data sets or causal models, and (c) assess the adequacy of explanations.
2. Revising (a) explanations if they have been found to be inadequate and (b) causal models in light of anomalous data.
3. Using models (a) to pose research questions and (b) to design data-generating activities as a part of answering questions.
4. Developing an appropriate technical language that can be used to converse about models and explanations in classrooms.
5. Determining what statements to accept about the world.
6. Considering (a) what reasoning patterns a discipline uses when conducting its inquiries and (b) what standards or norms are to be used to assess the quality of claims made in the classroom. (adapted from p. 13)

As apparent in this list, the heart of the science classroom involves the building and testing of models, argumentation, and the assessment of explanations.

In mathematics, the validation of propositions involves building logical proofs based on agreed-on common notions (e.g., definitions, axioms). To do this, students should first come to realize that mathematization of a concrete problem often involves creating a mathematical structure that represents one of the possible abstractions from the concrete situation. Validating the mathematical model created conforms to the list of proficiencies posed by Stewart et al. (2001), but students need to go beyond validating the mathematical model in terms of the concrete situation under study to assessing the consistency of the model itself in terms of its mathematical elements. Doing so involves having students develop "mini-deductive" systems within which generalizations can be "proved" in terms of some agreed-on notions, such as (in chap. 4) the justification of "zero added to another number equals that other number."

Students need to both comprehend and participate in mathematical and scientific argument, but as Forman (2001) noted, developing a classroom that supports student argumentation presents significant challenges for teachers and teacher educators:

> First, teachers must be able to understand students' explanations that may be inarticulate, incomplete, and quite different from each other in

real time. Second, teachers must be careful about silencing some students and privileging others. This is particularly difficult if the goal is to encourage students to share their thinking as well as to evaluate each other's strategies in terms of scientifically sound criteria (e.g., efficiency, precision, logical consistency, explicitness). Third, teachers need to help their students respect their classmates' explanations and criticisms and understand that learning requires identifying and profiting from one's own mistakes as well as those of others. Fourth, teachers need to communicate their instructional goals and practices to parents, other teachers, and administrators, who might not share their values and who are unlikely to have experienced this kind of instructional approach themselves. (pp. 10–11)

Regardless of the challenges faced, this type of learning environment more closely resembles the working environment of the scientist or mathematician, whether creating, describing, and defending a model; or conjecturing, making convincing arguments, and reflecting on his or her thinking. In classrooms built on such argumentation, students are acculturated into classroom communities of practice in which they create and internalize their own understanding of complex mathematical and scientific concepts through argumentation.

Classroom Norms for Developing Modeling and Argumentation

The norms in a particular class determine how students and the teacher are expected to act or respond to a particular situation. Normative practices are recurrent pairings of contexts and events in the classroom that signal to students expectations of what will happen and what they are to do, and they govern the nature of the models and arguments that students and teachers use to justify mathematical conjectures and conclusions. These norms can be made manifest through overt expectations or through more subtle messages that permeate the classroom environments.

Although the selection of appropriate tasks and tools can facilitate the development of modeling and argumentation, normative practices of a class determine whether they will be used for that purpose. In chapter 2, Lehrer and Schauble describe the importance of "activity structures" as recurrent pairings that teachers use to send signals to students about what counts as appropriate performance in contexts such as whole-class discussions, group work, and so on. They also serve to provide organizational structure to the sequence of daily activities. In classrooms with norms that support modeling and argumentation, tasks are viewed as opportunities to construct models and to engage in reasoned argumentation, not exercises to be completed using prescribed procedures. Similarly, tools are perceived as a means to represent models and to provide a context and focus for a discussion. In such classrooms, the process of learning is viewed as problem solving rather than as drill and practice, and students apply existing knowledge to generate new knowledge rather than to assimilate facts and procedures.

Classrooms can become discourse communities in which all students reason about alternative strategies or different ways of viewing important scientific or mathematical ideas. Students come to expect that the teacher and their peers want explanations for their conjectures and conclusions and for the validity of using a certain procedure to solve a given problem. In this way, mathematics and science become languages for thought rather than function as an acquired collection of facts and procedures that get correct, but sometimes inaccurate, answers.

Adopting normative practices related to developing modeling and argumentation requires a shift from norms common in traditional mathematics or science classrooms, in which the primary work of teachers has been to maintain order and control (Romberg & Carpenter, 1985). There is an inexorably logical sequence when the acknowledged work of teachers is to transmit concepts and skills. The most cost-effective way to transmit knowledge is through exposition to a captive audience, but that exposition cannot happen unless there is control, which is easier if students talk as little as possible and remain seated. This system for delivering knowledge to the group is essentially a system for controlling the individuals of that group. This simple sequence has dictated work, furniture arrangement, and school architecture for the last 100 years and is the tradition challenged by any attempt at change. Secada and Williams (chap. 11) describe the reluctance of many teachers to change instructional practices, and Webb, Romberg, Ford, and Burrill (chap. 10) document the difficulties several teachers encountered as they worked to adopt new practices.

This shift in teaching practices involves moving from a "mechanistic" perspective to an "environmental" one, in which teaching is based on the notion that the roles and work of student and teacher are complementary. The job of teaching, then, is to support, promote, encourage, and in every way facilitate student understanding. As teachers attempt to create a collaborative learning environment in which the students can explore and investigate ways to model problems and justify assertions and procedures, they guide, listen, discuss, prompt, question, and clarify the work students do. Teachers work to orchestrate appropriate activities, attend to each student's needs, and provide individual feedback on achievement. We note, however, that these norms are new to many teachers. As Little (1993) argued, this approach to mathematics and science teaching represents, "on the whole, a substantial departure from teachers' prior experience, established beliefs, and present practice. Indeed, they hold out an image of conditions of learning for children that their teachers have themselves rarely experienced" (p. 130).

To provide an environment that promotes deep reasoning, teachers need a deep understanding of the content appropriate to the instructional level at which they teach, including knowledge of (a) the mathematics or science being taught, (b) its relationship to the content prior to and beyond the instructional level of students, (c) the ways students reason about and come to understand that content at their instructional levels, and (d) the nature of mathematical and scientific practices. With this knowledge, teachers can competently guide their students through the twists and turns of rich mathematical or scientific discourse. In typical mathematics and science classrooms, unfortunately, many teachers at all school levels fail to have this level of understanding of mathematics or science. In chapter 9, Franke, Kazemi, Shih, Biagetti, and Battey show how a teacher's failure to understand student reasoning restricted students' strategies. In chapter 11, Secada and Williams discuss instances in which teachers and students thought that classroom discourse on alternate strategies was not legitimate classroom behavior.

When teachers shift to a focus on modeling and argumentation, the pattern of instruction also becomes less routine, which can pose a challenge, as Gail Burrill (quoted in de Lange, Burrill, Romberg, & van Reeuwijk, 1993) noted in her reflection on teaching a reform unit:

> There was little to "teach"; rather, the students had to read the map, read the keys, read the questions, determine what they were being asked to do, decide which piece of information from the map could be used to help them do this, and finally, decide what mathematics skills they needed, if any, in answering the question. There was no way the teacher could set the stage by demonstrating two examples (one of each kind), or by assigning five "seat work" problems and then turning students loose on their homework with a model firmly (for the moment) in place. (p. 154)

The changes in instruction forced teachers to reconsider how they interacted with their students.

Focusing on modeling and argumentation can also lead to new organizational relationships. In particular, teachers need to collaborate and share ideas. Chapters 2, 3, 4, 6, and 10 all address the issues of developing a professional community that involves shared planning time for teachers. As Rowan, Raudenbush, and Cheong (1993) argued, "Nonroutine forms of teaching lead to organic management because of the independent initiative of teachers who, in response to the challenges of nonroutine work, seek out and participate in organizational processes that help them cope more effectively with nonroutine work" (p. 485). In fact, Clarke (1993) found in his study that when teachers shared planning time, they "claimed that the support they received from each other, the opportunity to work together in planning and 'debriefing' each day, and the opportunity to have a 'sounding board' present (whether another teacher, a project staff member, or a researcher) all appeared to be facilitative of professional growth" (p. 224).

Learning as a Community Activity

Our conception of instruction in classrooms should be seen as a community activity. In the classrooms described in these studies, class activity often involved sharing models, arguments, and ideas with the goal of developing connections among them. To accomplish this, instruction focused on making classrooms discourse communities. Students shared their models, reasons, ideas, and strategies with other students in an open, supportive environment, and teachers learned not just to listen, *but also to hear*, what students said.

Such classes are engaged in practices of generating knowledge. Conjectures are proposed, and the members of the class often work together to refine and validate them. Often a number of students are involved in generating and refining a model. Artifacts adopted by the class become a basis for collective reflection and articulation of ideas. Some classes adopt a stance that knowledge generation is a community function and recognize that they do not have to depend on the teacher as the provider and arbitrator of what counts as knowledge. Not all mathematics or science is learned in classrooms, and teachers and students alike must recognize this. Students must be encouraged to make links between real-world and classroom mathematics and science. They should be encouraged to seek out applications of mathematics and science in their world and to continue their sense-making and construction outside the classroom.

CONCLUSION

In the United States, popular perception holds that our citizens need a better understanding of mathematics and science if our society is to prosper in this rapidly changing global economy. As a consequence, for nearly two decades mathematics education and science education have been the focus of public demand for action through changes in the way mathematics and science are taught and learned. The task of developing and demonstrating alternatives to current modes of mathematics and science instruction is a design task: Simple alterations of existing programs will not suffice. The fundamental way in which mathematics and science programs are organized and developed must be changed, and new programs created.

The conceptual basis for our work has been and continues to be centered on learning with understanding. Fennema and Romberg (1999) described in some depth our conceptions of what it means to learn with understanding and illustrated such learning in classrooms. In this chapter, we have attempted to refine this conception of learning with understanding and make explicit the relation between learning with understanding and the related constructs of modeling and argumentation. Properly integrated into instruction, modeling can provide students a specific context for engaging in reasoned argumentation. More than a valuable learning tool, reasoned argumentation introduces students to the experience of professional mathematicians and scientists, who likewise share and develop their ideas with colleagues through argumentation. Explanation and justification are at the heart of reasoned argumentation, but argumentation also involves reasoning by analogy and drawing on relationships. Working together, students generate and validate new concepts and procedures. As students engage in modeling and argumentation, they develop an identity that produces the confidence that they can generate models and engage in argumentation to make sense of science and mathematics.

In the classrooms we describe in this book, students and teachers worked to understand and engage in the practices of mathematics and science. They became reflective about the activities they engaged in as they explored new concepts, solved problems, and learned to look for relationships that gave meaning to new ideas. As they learned to critically examine their existing knowledge and to build new connections between concepts, they also came to view learning as (a) problem solving based in inquiry, (b) reasoning through models, and (c) argumentation sometimes characterized by conjecture and generally subject to validation.

REFERENCES

American Association for the Advancement of Science. (1993). *Benchmarks for science literacy: Project 2061*. New York: Oxford University Press.

Bazerman, C. (1988). *Shaping written knowledge: The genre and activity of the experimental article in science*. Madison, WI: University of Wisconsin Press.

Brown, A. (1992). Design experiments: Theoretical and methodological challenges in creating complex interventions in classrooms. *Journal of Learning Sciences, 2,* 141–178.

Brown, A., & Campione, J. (1996). Psychological theory and the design of innovative learning environments: On procedures, principles, and systems. In L. Schauble & R. Glaser (Eds.), *Innovations in learning: New environments for education* (pp. 289–325). Mahwah, NJ: Lawrence Erlbaum Associates.

Carey, S., Evans, R., Honda, M., Jay, E., & Unger, C. (1989). "An experiment is when you try it and see if it works": A study of Grade 7 students' understanding of the construction of scientific knowledge. *International Journal of Science Education, 11,* 514–529.

Carpenter, T. P., & Levi, L. (2000). *Developing conceptions of algebraic reasoning in the primary grades* (Res. Rep. No. 00-2). Madison, WI: National Center for Improving Student Learning and Achievement in Mathematics and Science, Wisconsin Center for Education Research.

Cartier, J., Rudolph, J., & Stewart, J. (2001). *The nature and structure of scientific models.* Madison, WI: National Center for Improving Student Learning and Achievement in Mathematics and Science, Wisconsin Center for Education Research.

Clarke, D. (1993). *Influences on the changing role of the mathematics teacher.* Unpublished doctoral dissertation, University of Wisconsin–Madison.

Cobb, P. (2001). Supporting the improvement of learning and teaching in social and institutional context. In S. Carver & D. Klahr (Eds.), *Cognition and instruction: Twenty-five years of progress.* Mahwah, NJ: Lawrence Erlbaum Associates.

Cobb, P., Boufi, A., McClain, K., & Whitenack, J. (1997). Reflective discourse and collective reflection. *Journal for Research in Mathematics Education, 23*(3), 258–277.

Cross, M., & Moscardini, A. O. (1985). *Learning the art of mathematical modeling.* Chichester, England: Ellis Horwood.

de Lange, J., Burrill, G., Romberg, T., & van Reeuwijk, M. (1993). *Learning and testing mathematics in context.* Pleasantville, NY: Wings for Learning.

Devlin, K. (1994). *Mathematics: The science of patterns.* New York: Freeman.

Fennema, E., & Romberg, T. A. (Eds.). (1999). *Classrooms that promote mathematical understanding.* Mahwah, NJ: Lawrence Erlbaum Associates.

Forman, E. A. (2001). *Classroom discourse project.* Madison, WI: National Center for Improving Student Learning and Achievement in Mathematics and Science, Wisconsin Center for Education Research.

Forman, E. A., Larreamendy-Joerns, J., Stein, M. K., & Brown, C. A. (1998). "You're going to want to find out which and prove it": Collective argumentation in a mathematics classroom. *Learning and Instruction, 8*(6), 527–548.

Freudenthal, H. (1983). *Didactical phenomenology of mathematical structures.* Dordrecht, Netherlands: Reidel.

Freudenthal, H. (1987). Mathematics starting and staying in reality. In I. Wirszup & R. Street (Eds.), *Proceedings of the USCMP conference on mathematics education on development in school mathematics education around the world* (pp. 279–295). Reston, VA: National Council of Teachers of Mathematics.

Gee, J. (1998). *Preamble to a literacy program.* Madison, WI: Department of Curriculum and Instruction, University of Wisconsin–Madison.

Goodwin, M. H., & Goodwin, C. (1987). Children's arguing. In S. U. Philips, S. Steele, & C. Tanz (Eds.), *Language, gender, and sex in comparative perspective* (pp. 200–249). Cambridge, England: Cambridge University Press.

Greeno, J. (1991). Number sense as situated knowing in a conceptual domain. *Journal for Research in Mathematics Education, 22*(3), 170–218.

Hestenes, D. (1992). Modeling games in the Newtonian world. *American Journal of Physics, 60,* 440–454.

Keitel, C. (1987). What are the goals of mathematics for all? *Journal of Curriculum Studies, 19*(5), 393–407.

Kilpatrick, J. (1987). George Polya's influence on mathematics education. *Mathematics Magazine, 60*(5), 299–300.

Kitcher, P. (1993). *The advancement of science: Science without legend, objectivity without illusions.* New York: Oxford University Press.

Kuyk, W. (1982). A neuropsychodynamical theory of mathematics learning. In *Proceedings of the sixth international conference for the Psychology of Mathematical Education* (pp. 48–62). Antwerp, Belgium: Universitaire Instelling.

Lehrer, R., & Schauble, L. (2003). Origins and evolution of model-based reasoning in mathematics and science. In R. Lesh (Ed.), *Models and modeling in mathematics teaching, learning, and problem solving* (pp. 59–70). Mahwah, NJ: Lawrence Erlbaum Associates.

Lesh, R., & Doerr, H. (in press). Foundations of models and modeling. In R. Lesh (Ed.), *Models and modeling in mathematics teaching, learning, and problem solving.* Mahwah, NJ: Lawrence Erlbaum Associates.

Little, J. (1993). Teachers' professional development in a climate of educational reform. *Educational Evaluation and Policy Analysis, 15*(2), 129–151.

McKnight, C., & Schmidt, W. (1998). Facing facts in U.S. science and mathematics education: Where we stand and where we want to go. *Journal of Science Education and Technology, 7*(1), 57–76.

National Commission on Excellence in Education. (1983). *A nation at risk: The imperative for educational reform.* Washington, DC: U.S. Department of Education, Office of Educational Research and Improvement, Center for Libraries and Education Improvement.

National Commission on Mathematics and Science Teaching for the 21st Century. (2000). *Before it's too late: A report to the nation.* Jessup, MD: Education Publications Center, U.S. Department of Education.

National Council of Teachers of Mathematics. (1989). *Curriculum and evaluation standards for school mathematics.* Reston, VA: Author.

National Council of Teachers of Mathematics. (1991). *Professional standards for teaching mathematics.* Reston, VA: Author.

National Council of Teachers of Mathematics. (1995). *Assessment standards for school mathematics.* Reston, VA: Author.

National Council of Teachers of Mathematics. (2000). *Principles and standards for school mathematics.* Reston, VA: Author.

National Research Council. (1996). *National science education standards.* Washington, DC: National Academy Press.

National Science Board Commission on Precollege Education in Mathematics, Science, and Technology. (1983). *Educating Americans for the 21st century: A plan of action for improving the mathematics, science, and technology education for all American elementary and secondary students so that their achievement is the best in the world by 1995.* Washington, DC: National Science Foundation.

No Child Left Behind Act of 2001. Pub. L. No. 107-110, 115 Stat. 1425 (2002).

Organization for Economic Cooperation and Development. (1999). *Measuring student knowledge and skills: A new framework for assessment.* Paris: Author.

Organization for Economic Cooperation and Development. (2001). *Knowledge and skills for life.* Paris: Author.

Organization for Economic Cooperation and Development. (2002). *Framework for the PISA mathematics domain.* Paris: Author.

Peak, L. (1996). *Pursuing excellence: A study of U.S. eighth-grade mathematics and science teaching, learning, curriculum, and achievement in an international context.* Washington, DC: National Center for Educational Statistics.

Penner, D. E., Giles, N. D., Lehrer, R., & Schauble, L. (1997). Building functional models: Designing an elbow. *Journal of Research in Science Teaching, 34,* 1–20.

Romberg, T. A. (1983). A common curriculum for mathematics. In G. D. Fenstermacher & J. I. Goodlad (Eds.), *Individual differences and the common curriculum* (pp. 121–159). Chicago: University of Chicago Press.

Romberg, T. A. (1992). Problematic features of the school mathematics curriculum. In P. Jackson (Ed.), *Handbook on research on curriculum* (pp. 749–788). New York: Macmillan.

Romberg, T. A., & Carpenter, T. P. (1985). Research on teaching and learning mathematics: Two disciplines of scientific inquiry. In M. Wittrock (Ed.), *The third handbook of research on teaching* (pp. 850–873). New York: Macmillan.

Rowan, B., Raudenbush, S. W., & Cheong, Y. F. (1993). Teaching as a nonroutine task: Implications for the management of schools. *Educational Administration Quarterly, 29*(4), 479–500.

Shafer, M., & Romberg, T. (1999). In E. Fennema & T. A. Romberg (Eds.), *Classrooms that promote mathematical understanding* (pp. 159–184). Mahwah, NJ: Lawrence Erlbaum Associates.

Silver, E. A., & Smith, M. S. (1996). Building discourse communities in mathematics classrooms: A worthwhile but challenging journey. In P. Elliott & M. Kenney (Eds.), *Communication in mathematics, K–12 and beyond* (pp. 20–28). Reston, VA: National Council of Teachers of Mathematics.

Simon, M. (1995). Reconstructing mathematics pedagogy from a constructivist perspective. *Journal for Research in Mathematics Education, 26,* 114–145.

Songer, N., & Linn, M. (1991). How do students' views of science influence knowledge integration. *Journal of Research in Science Teaching, 28,* 761–784.

Steen, L. A. (1990). *On the shoulders of giants: New approaches to numeracy.* Washington, DC: National Academy Press.

Stewart, J., Cartier, J., & Passmore, C. (2001). *Scientific practice as a context for inquiry and argumentation in science classrooms.* Madison, WI: National Center for Improving Student Learning and Achievement in Mathematics and Science, Wisconsin Center for Education Research.

van Reeuwijk, M., & Wijers, M. (2003). Explanations why?: The role of explanations in answers to (assessment) problems. In R. A. Lesh & H. Doerr (Eds.), *Beyond constructivism: Models and modeling perspectives on mathematics problem solving, learning, and teaching.* Mahwah, NJ: Lawrence Erlbaum Associates.

Webb, N., & Romberg, T. (1992). Implications of the NCTM standards for mathematics assessment. In T. Romberg (Ed.), *Mathematics assessment and evaluation* (pp. 37–60). Albany, NY: State University of New York Press.

Yackel, E., & Cobb, P. (1996). Sociomathematical norms, argumentation, and autonomy in mathematics. *Journal for Research in Mathematics Education, 27*(4), 458–477.

Yackel, E., Cobb, P., & Wood, T. (1991). Small-group interactions as a source of learning opportunities in second-grade mathematics. *Journal for Research in Mathematics Education, 22*(5), 390–408.

Part II

LEARNING WITH UNDERSTANDING

Chapter 2

Developing Modeling and Argument in the Elementary Grades

Richard Lehrer
Leona Schauble
Peabody College, Vanderbilt University

The models of nature that scientists construct are as diverse as their disciplines, and much of the rhetoric and practice of science is governed by efforts to invent, revise, and contest models. Our program of research was aimed at exploring the implications of this view of science in the education of children in the elementary grades: Can modeling practices be introduced to young children? Can teachers effectively support the growth and development of model-based reasoning? What forms of mathematics best support modeling practices?

Our research rested on a foundation of partnership with a public school district. The most visible aspect of the partnership centered on our efforts to support the development of a community of teacher-partners, who collectively fostered and studied the development of students' model-based reasoning. The research employed a design paradigm that featured the crafting of ecologies of learning to support the growth of model-based reasoning for teachers, students, researchers, and, in some senses, the school district as an institution.

Within these evolving contexts, we developed several related lines of research. First, we engaged with our teacher-partners in careful rethinking and empirical investigation of the forms of mathematics and science that could support model-based reasoning in the elementary grades. Second, we studied the teaching and institutional supports needed for the widespread adoption of these forms of mathematics and science, with an eye toward transforming practice, both within classrooms and at school-wide and district levels. Finally, we studied the forms of student learning that emerged as a result of these changes in institutional commitments and teaching practices. This chapter summarizes the findings from each of these lines of inquiry.

WHAT'S WORTH MODELING?
RETHINKING SCHOOL MATHEMATICS AND SCIENCE

We are committed to the proposition that students should have access to important and powerful ideas in mathematics and science at every grade. In most elementary schools, in contrast, instruction about these ideas is either neglected altogether or delayed until students have "mastered the basics." Yet it is well worth rethinking what should count as "the basics."

For example, instead of restricting young students' mathematical education to arithmetic computation, our teacher-partners and we broadened the picture to include the foundations of geometry and space, measurement, data and data structure, uncertainty, and early ideas about algebra. Of course, little is known about, for example, what geometry for primary students should look like, or about elementary students' capabilities for acquiring these ideas. Therefore, developing and testing appropriate "inroads" to new mathematical ideas has been a fundamental part of the enterprise (e.g., Lehrer & Chazan, 1998; Lehrer, Jacobson, Kemeny, & Strom, 1999; Lehrer & Romberg, 1996; Lehrer & Schauble, 2000b). In this body of work, we sought to challenge long-standing beliefs about the nature of mathematics education in the elementary grades. For example, with good instruction, very young students can meaningfully consider the epistemic grounds of generalization and proof (Lampert, 2001; Lehrer, Jacobson, Thoyre, et al., 1998; Lehrer & Lesh, 2003). Moreover, these epistemic considerations often arise when children investigate the mathematics of shape and form, measurement, and data.

Similarly, in the early grades, school science is most often either the study of a succession of unrelated topics (e.g., weather, rocks and minerals, the rain forest) or an attempt to impart a mythical set of content-free "skills" (often misdescribed as "the scientific method" or "scientific process skills"). Yet there is no reason why science instruction cannot build coherently and cumulatively around important core ideas, as described in the science standards (e.g., American Association for the Advancement of Science, 1993; National Research Council, 1996). Accomplishing this, however, will require the generation of a research base that can help teachers make meaningful connections between *what* is taught at the classroom level and *how*, over the course of a child's education, these topics build toward the development of student understanding of such themes as growth and diversity, structure and function, or behavior. In science and mathematics, therefore, we explored the teaching potential of these ideas and studied the conceptual resources and challenges that students bring to the learning of new forms of science. Our core agenda was to engage students in the progressive "mathematization" of nature (e.g., Kline, 1980; Lehrer & Schauble, 2000a, 2000c). For us, this is the heart of modeling, although we do not discount the vital role played by general familiarization with natural systems.

Modeling and Argument

Our emphasis on models follows from the widespread observation that, regardless of their domain or specialization, scientists' work involves building and refining models of the world (Giere, 1992; Stewart & Golubitsky, 1992). Core ideas like "diversity" or "structure" derive their power from the models that instantiate them, so to fulfill the promise of the "big ideas" outlined in national standards, students must realize these ideas as models. Our emphasis on argument follows from the widespread observation that models are not merely constructed but, equally important, are mobilized to support socially grounded arguments about the nature of physical reality (Bazerman, 1988; Latour & Woolgar, 1979; Pickering, 1995).

Of course, this approach entails the identification of forms of modeling that are well aligned with children's development, a problem that we have explored and revisited in a variety of realms (for more detail on this research, see Lehrer & Pritchard, 2003; Lehrer & Schauble, 2000c; Lehrer, Schauble, & Petrosino, 2001; Lehrer, Schauble, Strom, & Pligge, 2001; E. Penner & Lehrer, 2000). We learned that in children's instruction it is advisable to begin with models that resemble their target system (i.e., the phenomena being described or explained) in ways that can be easily detected, partly because resemblance helps children draw analogies between models and target systems (Brown, 1990). For example, when we gave students springs, wood, and assorted materials from a hardware store and asked them to construct a device that "works the way your elbow does," first graders' initial models were guided by perceptually salient correspondences (Grosslight, Unger, Jay, & Smith, 1991). In particular, most of the children used round foam balls to simulate the "bumps" in their elbow joints (D. Penner, Giles, Lehrer, & Schauble, 1997). This beginning concern with "looks like" eventually developed into a focus on "works like"—that is, a focus on relations among, and functions of, components of the target system (e.g., ways of constraining the motion of the elbow).

When students have space and geometry, measure, and data at their disposal, as well as the more traditional forms of number sense, the transition to mathematical modeling of natural phenomena becomes feasible and powerful, even (as noted earlier) in the early grades. Third graders, for example, investigated graphical and functional descriptions of the relationships between the position of a load and the point of attachment of the tendon, thus modeling elbows as third class levers (D. Penner, Lehrer, & Schauble, 1998).

To introduce students to modeling, we found it especially fruitful to begin with and then promote development of children's metarepresentational capabilities—their abilities to generate and selectively compose representations of phenomena (Cobb, Gravemeijer, Yackel, McClain, & Whitenack, 1997; diSessa, Hammer, Sherin, & Kolpakowski, 1991; Lehrer, Strom, & Confrey, 2002; Roth & McGinn, 1998). Following Latour (1990), we refer to these repertoires of representation—including diagrams, maps, drawings, graphs, texts, and related examples—as "inscriptions." Inscriptional competence is developed over years of schooling that emphasize generating, using, and progressively revising representations, so that core ideas come to be inscribed in multiple ways (Olson, 1994).

Modeling, of course, is a form of disciplinary argument that students learn over a long and extended period of practice and participation. In our consideration of what is worth thinking and learning, we pursued a general interest in children's appropriation of important disciplinary forms of argument. During their education, students should receive multiple opportunities to practice and acquire such forms of thinking, to understand when they are appropriate and the kinds of work they accomplish. In our own work, we also found interesting transitions in forms of argument within mathematics (Strom, Kemeny, Lehrer, & Forman, 2001). With good instruction, quite young students can shift from mathematical arguments based on particular cases to generalizations based on proof (Lehrer & Lesh, 2003). This shift in argument is accompanied by careful examination of epistemological foundations of mathematics, including the grounds of "knowing for sure," as one class of third graders described the enterprise. Finally, we encountered important differences between early epistemologies about mathematics and science. For example, a class of third graders was decidedly disconcerted by the fact that scientific models (e.g., models of the relationship between the volume and weight of objects composed of different materials) are only approximate, whereas similar relationships in the mathematical world are precise (Lehrer & Schauble, 2000a). In sum, our definition of what it means to understand includes both comprehending and participating in mathematical and scientific argument.

Long-Term Development

Because of our focus on core forms of reasoning and argument that are unlikely to be mastered within a unit or even a single school year, we emphasize students' long-term development of central conceptual and epistemic structures, rather than acquisition of nuggets of instruction "delivered" within brief periods. Decisions about curriculum and policy need to be guided by a big-picture view that not only encompasses short-term goals, but also gives a more encompassing sense of the purpose, direction, and desirable outcomes of a child's education. Strangely, schools often fail to entertain this kind of long-term view in their planning, scheduling, or assessment. For many reasons, teachers typically regard their work as bounded within their own grade of practice, and curriculum (because of student mobility and teacher independence of decision making) is often organized into interchangeable modules or units. Yet these practices do not conform well to a perspective of learning as a historical activity—one that assumes that current learning builds from and on learning achieved in earlier weeks, months, and years. Longer-term views of the development of student thinking need to undergird attempts to improve student achievement. Teachers need to become thoughtful about individual students' learning histories, and schools need to support conditions that permit learning to build progressively and to accumulate. Among other issues, decisions about what is taught need to be informed by this kind of long-term view.

To provide an example, we offer a brief snapshot of cross-grade modeling approaches to thinking about growth. Growth (of organisms and populations) and diversity, examples of unifying "big ideas" central to biology, afford entry via modeling both by young students and by students who are much older and more experienced. Our purpose in presenting this example is to show how science and mathematics learning can build systematically over years of instruction, offering opportunities for more sophisticated student inquiry and deeper understanding at each successive level. (For further details on the development of such ideas as thematic cores in elementary science, see Lehrer & Schauble, 2003.)

In our research, *first graders* represented the growth of flowering bulbs planted under different conditions (in soil, in water), using green paper strips to depict the height of plant stems at different points in the growth cycle (Lehrer, Carpenter, Schauble, & Putz, 2000). Depiction of height marked a transformation from thinking about the plant as whole to consideration of the plant as a set of attributes, height being most salient. Yet thinking about height provoked considerations of its measure, and considerations of measure reverberated to bring about reconsideration of the meaning of the attribute itself. Consistent with our findings that young children seek to preserve resemblance, students first insisted that the strips be colored green. However, as the teacher repeatedly focused students' attention on successive differences in the lengths of the strips by asking *how much faster* one plant grew than another, students began to make the transition from thinking of the strips as "presenting" height to "representing" height.

As they reasoned about changes in the differences of heights of the strips, children identified times when their plants grew "faster" or "slower." This identification relied on the arithmetic of comparative difference, a form of mathematics well within their grasp. They noted that the amaryllis bulbs grew faster at the beginning of their life cycle and then slowed, whereas the paperwhite narcissi grew very slowly at the beginning, but then "catched up." This work, of course, needed to be firmly grounded in prior discussions about what counted as "tall" and ways that attribute might be reliably measured. Building a firm understanding of the mathematics of measure was an important focus. We note that measure and attribute often co-originate at all grades and ages.

In the third grade, children worked to mathematize change in the growth of Wisconsin Fast Plants (Lehrer, Schauble, Carpenter, & Penner, 2000). They developed "pressed plant" silhouettes (records of changes in plant morphology over time), coordinate graphs that related plant height and time, sequences of rectangles representing the relationship between plant height and canopy "width," and three-dimensional forms that captured changes in plant volume. These forms of mathematization and inscription capitalized on children's mathematical understanding, especially their prior development of concepts of ratio in the context of geometric similarity (Lehrer, Strom, & Confrey, 2002).

As the diversity of students' representations increased, new cycles of inquiry emerged: Was the growth of roots and shoots the same or different? Comparing the height and depth of shoots and roots, students noticed that at any interval in the plant's life cycle, the differences in rates of growth were apparent. They also discovered, however, that graphs displaying the growth of roots and shoots were characterized by similar "shapes" (S-shaped logistic curves). Finding these likenesses in the *form* of the graphs but not in the *measure* of shoots and roots, students began to wonder about the significance of the observed similarity. Why might growth of different parts have the same form? When was growth the fastest, and what might be the functional significance of these periods of rapid growth? The variety of inscriptional forms that the third graders either invented or used provided many opportunities to develop metarepresentational competence. To provide just a single example, on one occasion students were puzzling over the apparent problem that coordinate graphs of two different plants looked similar (i.e., equally "steep"), yet actually represented different rates of growth. The mystery was solved when they discovered that the children who had generated the graphs used different scales to represent the height of their plants. The discovery that graphs might look "the same," yet represent different rates of growth, tempered the class' interpretations of coordinate graphs in other contexts throughout the year.

In the fifth grade, children again revisited ideas about growth, this time of tobacco hornworms (*Manduca sexta*). Their mathematical resources now included ideas about distribution and sample. Students explored relationships between growth factors (different food sources) and the relative dispersion of characteristics (length, width, time to pupation, survivorship) in the population at different points in the life cycle of the caterpillars. Students' questions focused on the diversity of characteristics within populations, not simply on shifts in central tendencies of attributes. As children's representational repertoires stretched, so did their considerations about what might be worthy of investigation (Lehrer & Schauble, 2002b).

In sum, over the span of the elementary school grades, we observed characteristic shifts from an early emphasis on literal depictional forms (Styrofoam balls in elbow models, paper "stems" in models of change in plant height) toward representations that were progressively more symbolic and mathematically powerful. Diversity in representation and mathematical resources both accompanied and produced conceptual change. As children developed a variety of mathematical means for characterizing growth, they came to understand biological change in more dynamic ways. For example, as students came to understand the mathematics of ratio and of changing ratios, they also began to conceive of growth not as a matter of simple increase, but as a patterned rate of change. These transitions opened up new avenues of inquiry: Was plant growth like animal growth (despite surface differences)? These forms of conceptual development did not simply occur, however; they required a context in which teachers systematically supported a related set of central ideas, building on earlier concepts to stretch students' understanding well beyond what is considered typical of most elementary students. Identifying mathematical and scientific models and concepts that can potentially serve as such a "core" and then working with teachers to develop the promise of these ideas across grades has been and continues to be one of the central strands of our research.

TEACHING AND INSTITUTIONAL SUPPORT FOR LEARNING

Although the principles described (e.g., curriculum that cumulates; emphasis on "big ideas" and model-based thinking; long-term developmental approaches to learning) were important constituents of our instructional design, their instantiation rested on our efforts to reconfigure the school community in ways that provided institutional support for innovation (and to cushion the impact of the inevitable failures along the way). Our primary focus was on teachers' professional development, an enterprise that we approached with an eye toward supporting teachers' development of knowledge—of mathematics, of science, of students' reasoning, and of the interrelationships. Although the need for support appeared self-evident, the means were not as clear, especially in light of the tentative and emerging quality of our collective understandings of model-based reasoning in the elementary grades.

Effective schools are more than places where outstanding teaching occurs. Because schools "organize human, technical, and social resources into a collective enterprise" (Newmann & Wehlage, 1995), we considered how to mesh support of individual teachers in that collective enterprise in ways that enabled them to participate in a wider, yet very tangible, community of practice (Borko & Putnam, 1996). Our efforts were informed by a dual perspective. One was a teacher's perspective, focusing on the forms of knowledge that might best inform the design of a classroom ecology to support model-based reasoning. The other was an institutional perspective, focusing on the ways professional development activity could be contextualized as an institutionally supported community. The latter goal involved negotiations with district administrators about format and venue for professional development, compensation for teachers, access to the district's existing financial resources for professional development, and the prospective fit between district goals (often encapsulated as standards for learning), and those of our research. In practice, these perspectives were interwoven. We regarded professional development not as a task to be accomplished within a bounded time period but rather, as an ongoing enterprise: Good teachers develop a commitment to continual learning—both their own and that of their colleagues—that deserves and requires sustained institutional and technical support.

Spadework Prior to Professional Development

The philosophy enunciated by the superintendent of this district (including two charter schools) was one of choice and competition among ideas and implementations. Our project was viewed as a new means to reach familiar ends (e.g., modeling practices to achieve standards in mathematics and in science). As we began our project, we built on a previous, smaller-scale collaboration with teachers of primary-grade mathematics who had worked together over the course of 3 years to develop a mathematics of space and geometry (Lehrer, Jacobson, Thoyre, et al., 1998). Participants in this research collaborative were partners in the conduct of design experiments, met regularly to share and analyze student work, authored texts for publication, and invented and shared a variety of tools and tasks for promoting the development of understanding (Lehrer & Curtis, 2000; Lehrer, Jacobson, et al., 1998; Nitabach & Lehrer, 1996; E. Penner & Lehrer, 2000). These participants became an important nucleus of teacher leadership, especially in the first years of our project. In addition, some of these teachers had taken the initiative to expand their reach to encompass all teachers in the primary grades at one of the schools. We regard this kind of effort to develop common values, goals, and practices as a mark of a generative community.

The leadership of the district perceived the previous project as making a tangible difference in mathematical teaching and learning. This had two very noticeable effects. First, the principal of the building most affected by the geometry research collaborative became an advocate of the modeling project. Her advocacy was informed by participating along with teachers in the professional development meetings and workshops, rather than by merely encouraging general sorts of reforms (e.g., "hands-on" science), as administrators typically do in most reform efforts (Spillane, 2000). Second, the director of curriculum re-purposed most of the professional development budget allocated to the elementary grades to support the modeling project. She was convinced that a continuation (and expansion) of the forms of professional development that were the hallmarks of the previous research collaborative would prove effective in a new effort. These financial and cultural-capital forms of district support lasted throughout the years of this research and served as important buffers during times of district transitions (see Gamoran et al., 2003). Nevertheless, important contradictions (which had implications for professional development) emerged between the activity of teachers and the norms and practices of district policies.

Conceptual Framework for Professional Development

Our support for teachers' professional development was informed by our view of teachers as designers of learning environments, and we intentionally organized our efforts around core themes (Lehrer & Schauble, 2000c). Consistent with our overall emphasis on modeling, we emphasized teachers' understanding of model-based reasoning in mathematics and science. Student reasoning was regarded as a form of mediated activity (Wertsch, 1998) that was contingent on a series of design features, including the tools that students either invented for themselves (e.g., a question that they developed) or that were posed by their teachers; the tools and inscriptions that they either invented or appropriated to accomplish these tasks; and the norms and means for making model-based arguments.

Model-Eliciting Tasks. We promoted teacher appropriation or invention of model-eliciting tasks (Lesh & Lammon, 1992), in which students created physical or mathematical descriptions of nature. We were especially interested in those that afforded easy entry for young or novice students, yet provided sufficient "lift" and complexity to challenge older or more experienced students and could be adapted to local circumstances. Moreover, we thought of the design of these tasks as cumulative, so that modeling cycles might extend over weeks, months, even years. We stressed that good modeling problems provoke variability in student thinking, an important resource that teachers can capitalize on by, for example, juxtaposing student positions so that they can be contrasted, even contested. Moreover, good problems provide feedback, which provides a press for model testing and model revision. The most useful tasks support a broad landscape of conceptual possibilities. Teacher goals, student interest, and other circumstances determine the pathway followed, but, ideally, other routes will also be open and, given a different set of agendas, alternative questions or forms of inquiry pursued.

Tools and Inscriptions. We also explored the ways that inscriptions and notations, including graphs, maps, diagrams, and algebraic formulae, can be used in conceptual development. Among younger and less experienced students, inscriptions serve as important precursors to fully developed modeling practices (Lehrer, Carpenter, et al., 2000; Lehrer & Pritchard, 2003). Participating teachers were encouraged to provide students with access to firsthand experiences with compelling or intriguing aspects of the world and then to support successive cycles of inscription and mathematization of these experiences. Inscriptional displays often raised new questions. For example, the changing rates of the growth of the heights of Wisconsin Fast Plants instigated new questions about whether or not root growth could be similarly characterized. These new questions, in turn, supported new cycles of inscription (Lehrer, Schauble et al., 2000). Hence, one goal of our professional development was to orient teachers toward supporting cycles of inquiry (anchored by inscriptions), which evolved over time, as mathematical and scientific investigations do.

Norms and Means for Model-Based Argumentation. We considered it essential for students not only to understand and evaluate the disciplinary arguments proposed by others, but also to develop and defend their own positions and conjectures using appropriate reasons and evidence and to explain their perspectives clearly. We worked with teachers to consider models as forms of argument (e.g., as tools for making claims visible) and to emphasize model competition. We do not consider these design features to be additive "factors," but mutually supportive aspects of classroom practice. Notations are associated with tasks that provide purpose and content for the need to represent with a notation, and they are typically developed to develop or communicate a particular model. Accordingly, teachers worked together to document and study how these features worked together to support the evolution of important forms of argument in their classrooms.

Supporting the Growth of a Learning Community

The work of teaching relies on participation in a community of practice, and our attempts to organize teaching around the design of learning environments highlighted the need to support the creation and growth of (what was for these teachers) a new form of community—one oriented toward learning and toward the long-term development of professional knowledge.

Teacher Institutes and Meetings. It is a cliché that learning takes time and effort, but seldom are the implications of that cliché brought to bear on the duration and frequency of opportunities for teachers to plan and learn together. For example, although our participating teachers were expected to spend increasing numbers of after-school hours working together, most of this time was devoted to fragmented initiatives (often administrative or policy dictates, almost never goals identified by the teachers themselves) seldom organized or sustained toward student learning. In contrast, we reorganized professional development around regular and more frequent occasions and forums for teachers to learn from each other. For example, teachers regularly spent blocks of time together each summer developing their disciplinary knowledge. Summer institutes, which typically lasted from 2 to 3 weeks, provided opportunities for teachers at the same level to participate in the kinds of mathematical and scientific modeling practices that their students might pursue during the school year. For instance, teachers' investigation of the mathematics of sampling and mapping was coordinated with a study of diversity of plant and animal life in the school's woodlot. Teachers studied the mathematics of change as they modeled the growth of a variety of species of trees; they explored the mathematics of classification while working with collection and classification of fossils. Teachers were offered the opportunity to enroll for university credit in these summer institutes, and the credits were applied by several of the teachers toward a professional master's degree program that we created, oriented toward the study of teaching and learning in schools.

In addition to the summer institutes, teachers met twice a month for half a day. We collaborated in the leadership of these meetings with teacher coordinators, who spent half their time teaching and half their time supporting the development of their colleagues. One of the monthly meetings was typically spent planning instruction (typically, in small cross-grade teams), working with project staff on target topics in math or science, or exploring intellectual and material resources (e.g., curricula, trade books, experts). The second meeting convened all the participating teachers. Meetings often focused group reflection on analyzing what a sample of student work revealed about that student's understanding. Groups of teachers read and discussed research digests prepared by the research staff concerning the development of student thinking. They also discussed grade-level assessments, parent math and science nights, and newsletters for parents. Teachers also occasionally requested that the research staff provide elective "informationals" about particular topics in mathematics and science (e.g., the geometry of triangles, elicitation and evaluation of children's questions, clustering strategies for multiplication, the development of place-value knowledge). Because we took a long-term view of prospective trajectories of teacher development, the program encouraged teachers to choose the format and degree of their participation. For example, during the first year, some teachers "tried on" only one lesson or topic in their classrooms, leaving more radical innovations of practice to succeeding years.

Teacher Authoring.   We also made time for teachers to work individually or in teams to develop modeling tasks and to document major transitions in student thinking as their students engaged in cycles of modeling. The documentation of student thinking was considered at least as important as the invention of classroom activities, and the cross-grade work ensured that the writing contributed to a developmental picture of long-term growth in student thinking. In these documents, teachers summarized their conclusions about student learning and supported these conclusions with evidence they observed in their classrooms. In creating these documents, teachers thought more deeply both about disciplinary knowledge and the ways students reasoned as they acquired this knowledge. Teachers' efforts were often organized as cases, which provided an important grounding in the ideas of development and a forum for teachers to organize their perspectives on the ways the design tools worked in concert to foster conceptual change. When teachers served as reviewers of each others' work, they also enacted a commitment to the learning of the group rather than merely to their own individual improvement. Teachers learned from each other as they reviewed analyses of student thinking or adapted instructional tasks or units invented by other colleagues. (For examples of the teachers' work, see Lehrer & Schauble, 2002a.)

Collaborations in the Classroom.   With teachers' collaboration and assistance, we conducted several studies of student learning in classrooms each year. The purpose of these design experiments (Cobb, Confrey, diSessa, Lehrer, & Schauble, 2003) was to focus more intensively on target topics of interest (e.g., children's generation and evaluation of classification systems, the use of geometric similarity as a model for density, studies of growth). These studies also had profound influences on teachers' professional development. In our yearly interviews, teachers reported that our questions and other forms of interaction in their classrooms served as important sources of reflection and, hence, of their own professional growth.

Promoting Professional Community. In fostering the development of teacher community, we initially attempted to break down grade-level barriers by encouraging teachers to think about the *long-term* development of "big ideas" (Schifter & Fosnot, 1993). We supported this change by making space for cross-grade planning, encouraging teachers to adapt instruction so as to teach "the same" ideas at different grade levels, and supporting the sharing of their analyses of student thinking. We also deliberately grouped together teachers from different grade levels to support and encourage their talking about the implications of what they were learning for students of different ages.

In addition, we generated forms of participation intended to enhance the professional aspects of teaching. We note, for example, that professionals take steps to further the profession as a whole rather than just honing individual private practice. They generate knowledge that contributes to a public and cumulative knowledge base, and they subject their claims to consensually adopted critical standards. Assuming these roles, however, is not necessarily consistent with the way society or many teachers themselves view their profession. We attempted to raise the likelihood that teachers would take on these new roles by sponsoring teacher-leadership positions, negotiating expectations about publication and review, and generating a variety of other forums for sharing professional and technical knowledge.

Documenting Change in Teachers' Knowledge and Practice

Documenting changes in teachers' knowledge and practices is no easy matter. There are often substantial gaps between "talking the talk" and "walking the walk" of reform (Spillane & Zeuli, 1999). Complicating this measurement problem was the variability in "uptake" of the program over the life of the project. There were fewer than 20 participants the first year and nearly 50 in the third and final year. At any point in time, the group represented wide variability on the continuum from "newcomer" to "old-timer" and in their intensity of participation. The district included the entire range, from "shadow teachers" (who never formally signed up but received considerable encouragement, resources, and advice from other participating teachers) to teacher-leaders (who were paid half-time to assist other teachers).We addressed these challenges by developing multiple, independent ways of tracking teacher change and seeking cross-validation: analysis of teacher documents, yearly teacher interviews, and case studies.

Teacher-Authored Documents. Changes in these documents over the years of the project provided insight into what teachers believed about productive instructional tasks, how they characterized student thinking, and what they offered as evidence for these claims. Naturally, these documents varied considerably in their subject-matter content and in the duration and intensity of the teaching and learning that they describe. Working initially with a subset of the papers and collaborating with Anna-Ruth Allen, we generated a category system that was intended to capture increasing attention to the details and the meaning of student thinking. *Unelaborated descriptions* were simple, straightforward descriptions of classroom activities, with few examples given of student talk or work and few reflections on student thinking. *Examples* gave descriptions of classroom events and examples of student talk or strategies but gave no analysis of, or commentary on, the examples. *Commentaries* contained comments or generalizations about, or assessment of, student understanding but no examples of student work or talk as illustration. *Analyses* included rich, detailed examples of student talk, thinking, or work, along with teacher reflections on students' strategies and thinking.

This system was subsequently applied to the entire group of papers. As Table 2.1 shows, teacher-authors became increasingly more reflective about student thinking over time. During the first year of analysis, over half fell into the first three categories. By contrast, of those collected during the third year, 83% included both examples and commentary or analysis, suggesting that the teacher-authors had indeed become more thoughtful at reflecting on student thinking.

We next classified the papers according to "genre." *Lesson plans* described objectives and activities but contained little or no analysis of student thinking or learning. *Narratives* were simply chronological descriptions (sometimes in considerable detail) of what occurred in the classroom, but little or no analysis or commentary. *Research papers* were organized around a central issue or question (e.g., students' understanding of ratio and proportion) and framed in relation to the teacher's investigation or argument.

TABLE 2.1
Teachers' Papers Coded for Attention to Student Thinking

Category	Year 1	Year 2	Year 3
		N (%*)	
1. Unelaborated description	–	1 (7%)	–
2. Example (little or no comment)	4 (44%)	1 (7%)	3 (17%)
3. Commentary (no examples)	1 (11%)	1 (7%)	–
4. Analysis (rich examples and teacher reflection)	4 (44%)	11 (79%)	15 (83%)
Total papers	9	14	18

*Does not total 100% because of rounding.

Narrative remained the predominant genre across the years of data collection (see Table 2.2). *Lesson plans*, however, declined steadily, and *research papers* increased over time to a third of those submitted. These changes reflected a shift toward explanation and inquiry, although most of the teachers apparently believed that a detailed account of what occurred was a satisfactory explanation.

Yearly Teacher Interviews. The yearly 3-hour teacher interviews involved two main sections: "Teacher Beliefs" (teachers' perceptions of peer collaboration, changes in professional identities and practices, and feedback on project meetings and resources) and "Student Thinking" (teachers' conjectures about what interview examples of children's classroom work might reveal about a given child's thinking).

The first section consisted of 32 questions. Some concerned matters of fact (e.g., "Have you made use of e-mail as a way to communicate or collaborate with other teachers this year?"); others were aimed directly at teacher conceptions (e.g., "We hear a lot about professional development nowadays. What does it mean for a teacher to develop or grow professionally?"). As we have reported elsewhere (Lehrer & Schauble, 2000c), the interviews reflected shifts in teachers' critical standards for effective practice, their views of community, and their conceptions of professional development: They signaled significant shifts in professional identity. One veteran participant, for example, explained that her initial focus on becoming an excellent teacher had widened to taking responsibility for the development of others:

> I think professional development means that you work with other professionals to develop ... if you got to the end of the year, and then sent your kids on [to another teacher who is more traditional], did you really make a difference? Now, I think the teachers here are learning what it means to become part of a community of learners, who learn how to collaborate and communicate and build knowledge together. You know, that's a whole lot more meaningful and worth a teacher's time and effort than that staff development thing, a workshop here or there, or a sprinkling of this, or buy-

TABLE 2.2
Teacher Papers Classified by Genre

Genre	Year 1 (N = 9)	Year 2 (N = 14)	Year 3 (N = 18)
Lesson plan	33%	21%	–
Narrative	56%	64%	67%
Research paper	11%	14%	33%

ing a new professional book, or whatever. It's really changing the way we as teachers work together. (Lehrer & Schauble, 2000c, p. 151)

Case Studies.   As the participant group swelled to nearly 50, staying abreast of changes in individual teachers' knowledge and practices became more difficult. For this reason, the research staff identified candidates who showed evidence of being on the cusp of an important shift in teaching practice. From this group, we then chose six who represented a range of experience with our program (from novices to "old-timers"), a range of grade levels (primary to upper elementary), and the four school buildings in the district. This subset of teacher-participants agreed to be observed on a regular schedule (at least once a week, but typically far more often, and at times several days in a row). The resulting case studies were intended to provide a more detailed picture of teacher change.

Observers kept detailed field notes of classroom activities and interactions, recorded and copied student work, and photographed classroom inscriptions. The case studies as a group revealed that changes in teachers' knowledge and practice occurred over an extended period. These changes were initially quite local: An insight that developed in one context or one corner of subject matter was not spontaneously transferred to other contexts (even those that seemed obviously related) or subject matters. Instead, these local colonies of competence gradually expanded outward and eventually hooked together (Giles, Steinthorsdottir, Fulton, Lehrer, & Schauble, 1999). One implication, consistent with findings from an experimental study conducted by Jacobson and Lehrer (2000), is that working to become an excellent teacher in one domain of mathematics does not necessarily or automatically result in becoming an excellent teacher in a different domain. Effective pedagogical knowledge and knowledge of student thinking are both tied tightly to domain.

The case studies also highlighted the importance of teachers receiving feedback from student talk and performance as their conjectures about student thinking develop. Over time, teachers began to seek out tasks, representations, and discussions that revealed student thinking.

District, Schools, and Classrooms

We began our work in a system in which an initial teacher community had begun to spread beyond its origins as a research collaborative. This engagement accelerated during the years of the project, and the district superintendent took to habitually "dropping by" during summer institutes to see what teachers were doing and to informally convey his approval, even his admiration, of their work. These informal meetings allowed teachers to engage in conversation about the "real" meaning of district standards and the like with the superintendent. We noted increasing evidence of teachers' recognition that they were developing the expertise to support teaching for understanding. (For discussion of the institutional structure and community politics that affected teaching and learning during this project, see Gamoran et al., 2003).

Although these forms of engagement tended to increase student achievement, they also produced strains as positions that were superficially aligned began to produce divergences in practice. The superintendent's embrace of choice, for example, created a space for our work, but as teachers worked hard to develop expertise, they came to see the district's embrace of choice as having an undesirable side effect, namely, endorsing programs in which ideology trumped evidence. Project teachers' increasing identification with the need to develop understanding and to document transitions in student thinking placed them at odds with teachers who continued to practice in more traditional ways. In one school, all teachers were asked to join the collaborative, and nearly all did. Eventually, too, their voices began to intrude on the exclusive franchise of the district leadership to set educational policy. These difficulties played out at both school and district contexts. The authority of curriculum coordinators was challenged. Teachers, for example, objected to policies that brought students out of their classrooms (e.g., "gifted and talented," "special needs") or that insisted on "direct instruction" for some students. Teacher-leaders within the community received support from the district (e.g., release time for coordinating the efforts of the group), but were not given institutionally sanctioned roles such as curriculum coordinator. These inconsistencies produced a tension between "lived" (e.g., Wenger, 1998) and formal roles of leadership, even as teacher-leaders began to take on work (e.g., scheduling meetings, setting agendas) previously reserved for administrators. Moreover, all schools had informal alliances of groups of teachers, some quite long standing. The community generated by the research sometimes stressed preexisting but unspoken school-wide norms (e.g., which teachers were perceived to be especially knowledgeable about mathematics and science), which were enforced by these alliances.

Second, the movement toward increased accountability, promoted by some members of the school board and by broader national movements, resulted in an increased reliance on state-mandated and related forms of standardized assessment. These forms of assessment are relatively easy to administer, and their outcomes are increasingly treated as indicators of student learning. Teachers initially felt constrained by the district's use of their students' assessment results as an evaluation of teacher effectiveness, but as their students met or (more often) exceeded state standards, teachers became concerned instead with the problem that these forms of assessment were not particularly informative about student thinking and, thus, provided little guidance for improved instruction. Nevertheless, their students' performance on these tests released teachers to turn their attention to more informative ways of assessing student knowledge.

Third, district and school administrators occasionally acted to hasten the spread of reform. At the end of the first year of our work, administrators assigned some of the participating teachers, who were new to the district, to another elementary school, in the hope that these teachers would prove catalysts for wider reform. Administrators seemed not to recognize the supportive role of the initial teaching community. But in the new schools, these teachers found themselves at variance with preexisting norms about the mathematics and science worth teaching and shifted into the role of leader, a change in role that some of the reassigned teachers were not ready to make. As a result, teachers with more experience in the program felt the need to increase their support of these newcomers, even though they believed that this effort detracted from the time they had to increase their own disciplinary and pedagogical knowledge.

The district also adopted new forms of mathematics curricula (e.g., the National Science Foundation–sponsored *Investigations in Number, Data, and Space*, TERC, 1995), but these efforts toward alignment again produced unexpected costs. As the district grew, with a corresponding influx of more new teachers, project teachers again felt compelled to spend inordinate amounts of time helping their new colleagues understand what these materials might offer for the design of instruction. Even some of the changes that we hoped for and supported, although eventually powerful and positive, sometimes generated unexpected "fallout" that had to be negotiated by all concerned.

As might be expected, parents and other members of the community also contributed to the overall landscape of change. On a biennial basis, they elected new school board members who reified national tensions. The decisions of the school board tended to vacillate from those supported by a narrow majority favoring more progressive forms of education to those backed by a minority who favored "back to basics" approaches and who occasionally bid for control of the board. To address these political challenges, our teacher-collaborators and we developed parent education programs and studied their impact on parent support of reform (Lehrer & Schauble, 2000c; Lehrer & Shumow, 1997). Winning the strong support of parents was essential to sustaining the changes in teaching and learning that the teachers were attempting to effect.

Student Learning and Achievement

Because students were learning forms of mathematics not typically taught in elementary grades (nor measured well or at all by current "age- or grade-appropriate" standardized assessments), we designed instruments to assess student achievement. Many of these items were adaptations from Lehrer's earlier research on geometry and measurement (Lehrer, Jacobson, et al., 1998). Our efforts were aimed at both widespread assessments and in-depth understanding of children's mathematical reasoning. Analysis of results of the achievement measures provided broad-scale views of growth in student learning over 3 years. Analysis of the interviews conducted with a sample of students tracked over 3 successive years provide insight into student understanding of the strategies they used and the concepts they learned.

The achievement items assessed student understanding of five related strands of mathematics: number, geometry, measure, early algebra, and data modeling and statistics. Most items were single-step problems for which students generated written responses. We included several released items from the National Assessment of Educational Progress (NAEP) to benchmark student achievement to national performance. After the first year, we constructed two forms of this instrument, one for Grades 1 and 2 (30 items) and the other for Grades 3 through 5 (53 items). Each year, forms were revised (but a core pool of items for all students retained) and then administered to all students at the end of school year.

Student responses were coded by pairs of independent raters (interrater reliability in excess of 98%). A 2-parameter (difficulty, discrimination) item-response theory (IRT) model was fit to each of the five forms of the achievement measure (one form for the first year and two forms, one for Grades 1 and 2 and one for Grades 3 through 5, for each of the following years). Consequently, each student's mathematical achievement could be expressed on a common scale. For ease of interpretation, we report results on an expected "proportion correct" scale.

Cross-Sectional Analyses

We first looked at yearly grade-level student achievement. Mean achievement scores by grade and year are displayed in Table 2.3. As might be expected, students in upper grades had higher achievement scores than those in lower grades. Looking at the data across years, we found that the greatest impact on student achievement at each grade level occurred during the second year of the project. Independent sample t-test results suggested that the differences between the first and second years were reliable at every grade ($p = .01$ for Grade 1; $p < .001$ for all remaining grades), with effect sizes of 0.56 (Grade 1), 0.94 (Grade 2), 0.43 (Grade 3), 0.54 (Grade 4), and 0.72 (Grade 5). All effect sizes indicate substantial gains in student achievement. Table 2.3 further suggests that these gains were maintained in the third year of the project.

TABLE 2.3
Scaled Achievement Scores: Means and Standard Deviations

	Year 1		Year 2		Year 3	
Grade			Mean (SD)			
1	25.0	(6.3)	32.5	(11.9)	—	—
2	30.1	(7.6)	43.3	(11.9)	47.5	(19.2)
3	39.3	(9.6)	46.7	(14.3)	48.2	(10.9)
4	41.7	(10.2)	51.2	(14.6)	50.7	(11.4)
5	47.7	(7.9)	62.2	(18.5)	59.0	(13.3)

Note. Because of teacher mobility, only small samples were available for the first-grade students in the third year.

Longitudinal Analyses

IRT estimates of latent ability provided a basis for examining growth in individual student achievement from year to year. For ease of interpretation, the equated ability estimates are again reported on an "expected proportion correct" scale. Growth was measured as a change in expected score on the 1998 Grades 3–5 form. Table 2.4 displays average gain scores for students who participated in the project from 1996 to 1997 and from 1997 to 1998. Also indicated is the proportion of students who demonstrated positive growth over this time period. The average gain scores indicate substantial growth in individual student achievement, and the proportions demonstrate that achievement gains were widespread.

Comparison to National Samples

We developed rough benchmarks of comparison to national achievement by selecting seven NAEP-released items related to the mathematical

TABLE 2.4
Longitudinal Gains in Scores From Previous Year:
Means and Standard Deviations

	Longitudinal gains from previous year		Proportion of students who demonstrated positive growth
Year	Mean (SD)		
Year 2 (N = 96)	17.64	(9.65)	0.94
Year 3 (N = 183)	11.04	(14.59)	0.80

strands assessed in our instrument. Student performance in the second year of the project (1997) was then compared to average national performance on the NAEP for each item. Performance of participating students compared very favorably to national levels of performance for all items, although definitive conclusions are not possible because of differences in sample characteristics, stakes of the assessment, and other related factors. Typically, participating Grades 1 and 2 students performed very well on items that students in Grades 3 and 4 in the national sample found difficult; participating Grades 3–5 students outperformed Grades 8 and 12 students in the national sample. These comparisons, along with continued high performance on the standardized tests normally given by the school district, confirmed for parents and other stakeholders that student learning was not suffering under these innovations.

Summary of Achievement Assessment

Comparisons of student performance to national samples on selected NAEP items indicated that participating students' performances were comparable to those of students more advanced in grade level and age. These comparisons cannot be used as sole indicators of program effects, but they suggest that participating children achieved in a manner consistent with our expectations. Second, using the first year as a "control," average student achievement increased substantially at every grade level. Moreover, our estimate of comparative improvement in student achievement was conservative; student knowledge about mathematics at the end of the first year was probably already greater than that resulting from traditional content emphases and traditional practices. However, in light of the fact that it made little sense to conduct comparisons to other students in the district who were not learning these concepts, the year-to-year comparison is our best index of improvement. Third, gains in achievement were made by an overwhelming majority of students, not just those of high ability.

CONCLUSIONS

In this chapter, we summarized our findings in three primary areas of our work. First, we attempted to reformulate an approach to mathematics and science that honors both the history and epistemology of the disciplines *and* the resources that elementary school children bring to the learning enterprise. Second, we worked intensively to instigate and support teachers' working affiliation as a professional community, organized around the study and improvement of student thinking. Our research establishes that these efforts resulted in changes in teachers' beliefs and practices. Although our most intensive efforts were devoted to professional development, we also found it necessary to extend our educational thinking more broadly to school administration and organization (both formal and informal) and to the surrounding parents and community. Third and finally, we documented impressive cross-sectional and longitudinal gains in achievement among our participating students.

Perhaps the most important lesson from this work is that studying these factors in isolation, important though they are, makes little sense: The interconnections are evident. Focus on the design of learning environments turns attention to what is worth teaching and how best to teach that content, but these analyses are of little use unless teachers are both willing and able to learn how to teach more challenging forms of mathematics and science than they themselves were probably offered as students. Even a talented and willing body of teachers can be disheartened or derailed by an administrative structure with inconsistent values, or perhaps merely a lack of imagination about ways to support and sustain excellence. Changing professional roles offer new opportunities, but can also mean unacceptable transition to those who have a large stake in existing structures and affiliations. Parents and communities must be kept informed if we are to avoid the kind of backlash that dismantled the post-Sputnik education reform efforts.

Negotiating educational decisions in ways that acknowledge the complexities and interrelationships among these issues is important. We adopted a design philosophy that focused on classrooms and then moved to generate cross-grade and cross-school collaborations. This framework was used to develop instructional leadership and community among teachers and, where possible, to embrace curriculum coordinators, principals, and superintendents. Reform efforts interwove these different communities and were sustained by teachers' growing technical knowledge about what might be worth teaching and how students learn.

Our evidence shows that this approach to instruction provided children greater opportunity to learn significant mathematics and science. We look forward to continued conversation about how the trade-offs we have made and what we have learned as a result can contribute to a broader body of knowledge about the interdependencies of classroom design, professional development, student learning, institutional support, and equity.

REFERENCES

American Association for the Advancement of Science. (1993). *Benchmarks for science literacy*. New York: Oxford University Press.

Bazerman, C. (1988). *Shaping written knowledge: The genre and activity of the experimental article in science*. Madison: University of Wisconsin Press.

Borko, H., & Putnam, R. (1996). Learning to teach. In R. C. Calfee & D. C. Berliner (Eds.), *Handbook of educational psychology* (pp. 673–708). New York: Simon & Schuster.

Brown, A. L. (1990). Domain-specific principles affect learning and transfer in children. *Cognitive Science, 14,* 107–133.

Cobb, P., Confrey, J., diSessa, A., Lehrer, R., & Schauble, L. (2003). Design experiments in education research. *Educational Researcher, 32*(1), 9–13.

Cobb, P., Gravemeijer, K., Yackel, E., McClain, K., & Whitenack, J. (1997). Mathematizing and symbolizing: The emergence of chains of signification in one first-grade classroom. In D. Kirshner & J. A. Whitson (Eds.), *Situated cognition theory: Social, semiotic, and neurological perspectives* (pp. 151–233). Mahwah, NJ: Lawrence Erlbaum Associates.

diSessa, A., Hammer, D., Sherin, B., & Kolpakowski, T. (1991). Inventing graphing: Meta-representational expertise in children. *Journal of Mathematical Behavior, 10,* 117–160.

Gamoran, A., Anderson, C. W., Quiroz, P. A., Secada, W. G., Williams, T., & Ashmann, S. (2003). *Transforming teaching in math and science: How schools and districts can support change.* New York: Teachers College Press.

Giere, R. N. (Ed.). (1992). *Minnesota studies in the philosophy of science: Vol. 15. Cognitive models of science.* Minneapolis: University of Minnesota Press.

Giles, N. D., Steinthorsdottir, O., Fulton, D., Lehrer, R., & Schauble, L. (1999, April). *The transformation to teaching math for understanding in six elementary classrooms.* Paper presented at the annual meeting of the American Educational Research Association, Montreal, Quebec, Canada.

Grosslight, L., Unger, C., Jay, E., & Smith, C. (1991). Understanding models and their use in science: Conceptions of middle and high school students and experts. *Journal of Research in Science Teaching, 28,* 799–822.

Jacobson, C., & Lehrer, R. (2000). Teacher appropriation and student learning of geometry through design. *Journal for Research in Mathematics Education, 31*(1), 71–88.

Kline, M. (1980). *Mathematics: The loss of certainty.* Oxford, England: Oxford University Press.

Lampert, M. (2001). *Teaching problems and the problems of teaching.* New Haven, CT: Yale University Press.

Latour, B. (1990). Drawing things together. In M. Lynch & S. Woolgar (Eds.), *Representation in scientific practice* (pp. 19–68). Cambridge, MA: MIT Press.

Latour, B., & Woolgar, S. (1979). *Laboratory life: The construction of scientific facts.* Princeton, NJ: Princeton University Press.

Lehrer, R., Carpenter, S., Schauble, L., & Putz, A. (2000). Designing classrooms that support inquiry. In J. Minstrell & E. van Zee (Eds.), *Teaching in the inquiry-based science classroom* (pp. 80–99). Reston, VA: American Association for the Advancement of Science.

Lehrer, R., & Chazan, D. (Eds.). (1998). *Designing learning environments for developing understanding of geometry and space.* Mahwah, NJ: Lawrence Erlbaum Associates.

Lehrer, R., & Curtis, C. L. (2000). Why are some solids perfect? Conjectures and experiments by third graders. *Teaching Children Mathematics, 6,* 324–329.

Lehrer, R., Jacobson, C., Kemeny, V., & Strom, D. (1999). Building on children's intuitions to develop mathematical understanding of space. In E. Fennema & T. A. Romberg (Eds.), *Mathematics classrooms that promote understanding* (pp. 63–87). Mahwah, NJ: Lawrence Erlbaum Associates.

Lehrer, R., Jacobson, C., Thoyre, G., Kemeny, V., Danneker, D., Horvath, J., Gance, S., & Koehler, M. (1998). Developing understanding of space and geometry in the primary grades. In R. Lehrer & D. Chazan (Eds.), *Designing learning environments for developing understand of geometry and space* (pp. 169–200). Mahwah, NJ: Lawrence Erlbaum Associates.

Lehrer, R., & Lesh, R. (2003). Mathematical learning. In W. Reynolds & G. Miller (Eds.), *Handbook of psychology: Vol. 7. Educational psychology* (pp. 357–391). New York: Wiley.

Lehrer, R., & Pritchard, C. (2003). Symbolizing space into being. In K. Gravemeijer, R. Lehrer, L. Verschaffel, & B. Van Oers (Eds.), *Symbolizing, modeling, and tool use in mathematics education* (pp. 59–86). Dordrecht, Netherlands: Kluwer.

Lehrer, R., & Romberg, T. (1996). Exploring children's data modeling. *Cognition and Instruction, 14,* 69–108.

Lehrer, R., & Schauble, L. (2000a). Developing model-based reasoning in mathematics and science. *Journal of Applied Developmental Psychology, 21*(1), 39–48.

Lehrer, R., & Schauble, L. (2000b). Inventing data structures for representational purposes: Elementary grade students' classification models. *Mathematical Thinking and Learning, 2,* 51–74.

Lehrer, R., & Schauble, L. (2000c). Modeling in mathematics and science. In R. Glaser (Ed.), *Advances in instructional psychology: Vol. 5. Educational design and cognitive science* (pp. 101–159). Mahwah, NJ: Lawrence Erlbaum Associates.

Lehrer, R., & Schauble, L. (2002a). *Investigating real data in the classroom: Expanding children's understanding of math and science.* New York: Teachers College Press.

Lehrer, R., & Schauble, L. (2002b). Symbolic communication in mathematics and science: Co-constituting inscription and thought. In E. D. Amsel & J. Byrnes (Eds.), *The development of symbolic communication* (pp. 167–192). Mahwah, NJ: Lawrence Erlbaum Associates.

Lehrer, R., & Schauble, L. (2003). Origins and evolution of model-based reasoning in mathematics and science. In H. Doerr & R. Lesh (Eds.), *Beyond constructivism: A models and modeling perspective* (pp. 59–70). Mahwah: NJ: Lawrence Erlbaum Associates.

Lehrer, R., Schauble, L., Carpenter, S., & Penner, D. (2000). The interrelated development of inscriptions and conceptual understanding. In P. Cobb, E. Yackel, & K. McClain (Eds.), *Symbolizing and communicating in mathematics classrooms: Perspectives on discourse, tools, and instructional design* (pp. 325–360). Mahwah, NJ: Lawrence Erlbaum Associates.

Lehrer, R., Schauble, L., & Petrosino, A. (2001). Reconsidering the role of experiment in science education. In K. Crowley, C. D. Schunn, & T. Okada (Eds.), *Designing for science: Implications from everyday, classroom, and professional settings* (pp. 251–278). Mahwah, NJ: Lawrence Erlbaum Associates.

Lehrer, R., Schauble, L., Strom, D., & Pligge, M. (2001). Similarity of form and substance: Modeling material kind. In S. M. Carver & D. Klahr (Eds.), *Cognition and instruction: Twenty-five years of progress* (pp. 39–74). Mahwah, NJ: Lawrence Erlbaum Associates.

Lehrer, R., & Shumow, L. (1997). Aligning the construction zones of parents and teachers for mathematics reform. *Cognition and Instruction, 15,* 41–83.

Lehrer, R., Strom, D., & Confrey, J. (2002). Grounding metaphors and inscriptional resonance: Children's emerging understanding of mathematical similarity. *Cognition and Instruction, 20,* 359–398.

Lesh, R., & Lammon, S. J. (1992). Trends, goals, and priorities in mathematics assessment. In R. A. Lesh & S. J. Lammon (Eds.), *Assessment of authentic performance in school mathematics* (pp. 3–62). Washington, DC: American Association for the Advancement of Science.

National Research Council. (1996). *National science education standards.* Washington, DC: National Academy Press.

Newmann, F. M., & Wehlage, G. G. (1995). *Successful school restructuring.* Madison: Wisconsin Center for Education Research, University of Wisconsin–Madison.

Nitabach, E., & Lehrer, R. (1996). Developing spatial sense through area measurement. *Teaching Children Mathematics, 8,* 473–476.

Olson, D. R. (1994). *The world on paper: The conceptual and cognitive implications of writing and reading.* New York: Cambridge University Press.

Penner, D., Giles, N. D., Lehrer, R., & Schauble, L. (1997). Building functional models: Designing an elbow. *Journal of Research in Science Teaching, 34,* 125–143.

Penner, D., Lehrer, R., & Schauble, L. (1998). From physical models to biomechanics: A design-based modeling approach. *Journal of the Learning Sciences, 7,* 429–449.

Penner, E., & Lehrer, R. (2000). The shape of fairness. *Teaching Children Mathematics, 7*(4), 210–214.

Pickering, A. (1995). *The mangle of practice: Time, agency, and science.* Chicago: University of Chicago Press.

Roth, W. M., & McGinn, M. K. (1998). Inscriptions: Toward a theory of representing as social practice. *Review of Educational Research, 68,* 35–59.

Rudolph, J. L., & Stewart, J. (1998). Evolution and the nature of science: On the historical discord and its implications for education. *Journal of Research in Science Teaching, 35,* 1069–1089.

Schifter, D., & Fosnot, C. T. (1993). *Reconstructing mathematics education.* New York: Teachers College Press.

Spillane, J. (2000). Cognition and policy implementation: District policymakers and the reform of mathematics education. *Cognition and Instruction, 18,* 141–179.

Spillane, J. P., & Zeuli, J. S. (1999). Reform and teaching: Exploring patterns of practice in the context of national and state mathematics reforms. *Educational Evaluation and Policy Analysis, 21,* 1–27.

Stewart, I., & Golubitsky, M. (1992). *Fearful symmetry: Is God a geometer?* London: Penguin.

Strom, D., Kemeny, V., Lehrer, R., & Forman, E. (2001). Visualizing the emergent structure of children's mathematical argument. *Cognitive Science, 25,* 733–773.

TERC. (1995). *Investigations in number, data, and space.* Palo Alto, CA: Seymour.

Wenger, E. (1998). *Communities of practice: Learning, meaning, and identity.* Cambridge, England: Cambridge University Press.

Wertsch, J. V. (1998). *Mind as action.* New York: Oxford University Press.

Chapter 3

The Generative Potential
of Students' Everyday Knowledge
in Learning Science

Ann S. Rosebery
Beth Warren
Cindy Ballenger
Mark Ogonowski
TERC

Are children's everyday ways of knowing and talking a generative resource in learning science? If so, how? At least two traditions have emerged within the field of science that revolve around the relationship between "everyday" and "scientific" knowledge and knowing. In one, this relationship is viewed as dichotomous—descriptive of differences in the knowledge, knowing, and language use characteristic of scientists and nonscientists. Children's everyday ideas and ways of knowing and talking are seen as largely different from and incompatible with those of science. Research on misconceptions (e.g., Clement, 1982; McCloskey, Caramazza, & Green, 1980; McDermott, Rosenquist, & van Zee, 1987), for example, holds that students' everyday ideas are strongly held, can interfere with learning, and need to be replaced with correct conceptions. Within this tradition, everyday experience is viewed as problematic, as something to be overcome by properly designed science education.

In the second tradition, the relationship between these two forms of knowing is conceived as continuous. Research in this tradition focuses on understanding and characterizing the productive conceptual, metarepresentational, linguistic, experiential, and epistemological resources students have for advancing their understanding of scientific ideas (Ballenger, 1997; Clement, Brown, & Zeitsman, 1989; diSessa, 1993; diSessa, Hammer, Sherin, & Kolpakowski, 1991; Hammer, 2000; Hudicourt-Barnes, 2003; Lehrer & Schauble, 2000; Minstrell, 1989; Nemirovsky, Tierney, & Wright, 1998; Smith, diSessa, & Roschelle, 1993; Warren, Ballenger, Ogonowski, Rosebery, & Hudicourt-Barnes, 2001; Warren & Rosebery, 1996). This work does not assume a simple isomorphism between what children do and what scientists do; rather, it views the relationship as complex and taking a variety of forms: similarity, difference, complementarity, and generalization. In this tradition, science learning is viewed as a process of refinement and elaboration of everyday intuitions, modes of arguing, and experience.

When applied to the activity of children from groups that are historically underrepresented in the sciences, these traditions carry different assumptions about students' learning that, in turn, result in different implications for teaching. In the dichotomous view, children from historically underrepresented groups appear disadvantaged because their everyday ways of knowing and talking are often characterized as being the furthest from those traditionally valued in school science or in national standards. Recent studies of cultural congruence (Lee & Fradd, 1998; Lee, Fradd, & Sutman, 1995), for example, identify what the authors defined as "incompatibility" between the habits of mind and language and other interactional practices of students from certain language-minority groups (e.g., Haitian-Creole and Spanish-speaking communities) and those valued in national science standards. They suggest that these habits of mind and interactional practices can impede students' learning in science. From this point of view, differences between the sociocultural, cognitive, and linguistic practices of particular linguistic or ethnic groups and those of "mainstream" (e.g., white, middle-class) groups are conceptualized as potential barriers to understanding.

Studies in the second tradition have documented the ways in which the ideas and ways of talking and knowing of children from historically underrepresented groups are productively related to those characteristic of scientific communities (Ballenger, 1997, 2000; Gee & Clinton, 2000; Michaels & Sohmer, 2000; Warren et al., 2001; Warren & Rosebery, 1996). In our work in bilingual and economically and ethnically heterogeneous classrooms, for example, we view scientific sense-making as encompassing practices widely recognized as important in science (e.g., observing phenomena, collecting and analyzing data, building generalizations) as well as those less well known (e.g., arguing classroom evidence, imagining oneself inside a scientific phenomenon, re-presenting everyday experience).

The social studies of science literature reveals that scientists themselves engage in a wide range of ways of talking, knowing, and acting. Research shows that science is fundamentally a socially situated, heterogeneous practice that draws on diverse conceptual, symbolic, discursive, material, experiential, and imaginative resources (Keller, 1983; Latour, 1986; Latour & Woolgar, 1986; Ochs, Gonzales, & Jacoby, 1996; Root-Bernstein, 1989; Salk, 1983; Wolpert & Richards, 1997). By documenting and characterizing the everyday work and talk of scientists, these studies reveal a greatly expanded view of scientific practice that goes beyond emphasis on hypothetico-deductive reasoning and theory building, everyday experience as a form of misconception, and informal language as inadequate to the task of precise description, explanation, and modeling. These studies both demystify and deepen our sense of knowledge-making in science, showing it to be more heterogeneous and socially situated than standards-based descriptions of it suggest.

In light of these traditions, we decided to explore everyday and scientific ideas and practices in relation to (a) what children actually *do* when they are learning science and (b) our own assumptions, rather than to continue to conceptualize everyday and scientific as distinct domains. To do this, we formed a research partnership with elementary and middle school teachers in the Cambridge and Boston (MA) Public Schools to design new pedagogical approaches to improve science education for all students, including those from groups that are historically underrepresented in the sciences. In the sections that follow, we examine three cases in which children's everyday ways of knowing and talking figured prominently in their learning.

CASE 1: RE-PRESENTING EVERYDAY EXPERIENCE

Despite a strong overall commitment to constructivist theory, science education research has in very few cases focused on documenting, in the felicitous phrase of diSessa et al. (1991), the "substantial expertise" that children bring to the study of science. In this study of first- and second-grade students' learning about motion, we worked to better understand how children used their everyday experience of motion down an incline (e.g., bikes speeding down hills, their bodies moving down ramps) to make sense of acceleration due to gravity, in this instance, the changes in a toy car's speed as it rolled down a ramp.

In our analysis, we use "seeing" as a metaphor (see Goodwin, 1994, 2000; Latour, 1986; Ueno, 2000) for what is entailed in learning complex scientific ideas, in this case, the teacher's and the children's work to *make visible* what is impossible to observe directly: a toy car's positively and continuously changing speed as it moves down a ramp. Students were *learning to see* motion on a ramp from a Newtonian perspective, one that itself had to be developed through their joint work. How did this group of learners and their teacher make visible, in a way accountable to themselves and to the discipline of physics, the continuously changing speed of a toy car as it moved down a ramp?

The Study

Students in this combined first- and second-grade classroom investigated the motion of a toy car down a ramp as part of a larger investigation into motion involving acceleration, gravity, and inertia. The teacher, Suzanne Pothier, was interested in exploring what her students would make of the question, "What makes a toy car go down a ramp?" Ms. Pothier collaborated with researchers from the Chèche Konnen Center to significantly adapt the district's approved curriculum unit, *Insights: Balls and Ramps* (Education Development Center, 1997), to focus on visible and invisible forces. During the study, Ms. Pothier and researchers met weekly to study videotape and transcripts of the children's talk and activity. Classroom activities based on the children's questions and ideas were designed at these meetings.

The children explored pushes and pulls that they could directly observe or feel, for example, experimenting with ways to get a toy car moving from rest. One boy noticed that at times the car seemed to "move by itself," as when it rolled down his arm or was shaken inside a plastic pencil box. This kind of motion, "moving by itself," became the subject of intense investigation involving iterative cycles of experimentation, observation, description, and explanation, which were designed to engage the children in learning to "see" and to account for changes in the toy car's speed as it moved down an incline.

Prior to the focal episode (the second lesson in a series on observing and describing changes in speed as the car moved down a ramp), Ms. Pothier introduced the children to Isaac Newton and to some of his ideas about motion. Her intention was to put on the table Newton's perspective on the kind of motion they were investigating, along with toy cars, ramps, and other tools for investigating the car's motion. Ms. Pothier shared with the children a version of Newton's First and Second Laws, namely, that if a *change* occurred in the speed of an object or in the direction an object was moving, something had to be forcing that to happen. She then asked them to work in pairs to observe a toy car as it moved down a short ramp, to note any changes in speed they observed, and to create a story of the car's trip in terms of changes in speed. This proved immensely generative: Students not only noticed changes in the toy car's speed but also began to wonder about what might be "forcing" these changes.

The Focal Episode

This episode occurred about 17 minutes into the class. After one pair of students described how they thought their car "started to get faster" in the middle and reached its full speed near the end of the ramp, Elton suggested that the car's motion down the ramp was "sort of the same thing" as running down the ramp at school:

Elton: Putting the car down, down the ramp [points to a ramp] is sort of the same as you running down the ramp [makes quick waving motion] because [pauses] (not only) if you're in a car you go down a ramp [quickly angles hand from left to right] but also when you run down the ramp you could feel getting— when you're running faster [walks fingers at a descending angle from right to left] (... getting faster) [arcs hand right to left at eye level].

SP: So you've noticed your experience running down a ramp. How many people have run down a ramp outside? You know the ramp to go to the playground? [Many hands go up.] So tell us again what you've noticed, Elton.

Elton: Um, if a—if a car—if a car could get faster going down a ramp [angles hand down from right to left] you could because, um, um, when you get down and down [repeats gesture] it makes your legs go faster [rolls his hands] um—

SP: Mm, what did other people notice when you're running down a ramp? Do you go faster when you're running down a ramp?

In this segment, Elton's everyday experience figured in his efforts to make visible the relations between different instances of motion down a ramp. Elton clearly drew an analogy between the car's trip and his own experience running down a ramp. He then went on to explain how these two motions were the same. The critical relation he identified was one's feeling as the car one is in or one's body *goes* or *gets faster*. In explaining this, he not only imagined himself into a car but also recounted what it was like if "you" ran down a ramp ("you could feel getting—when you're running faster ... getting faster"). He engaged in what we have elsewhere called "embodied imagining" (Warren et al., 2001). He enacted his explanation with gestures of hands and fingers, imagining these scenes through his body and thinking with his body about what made them similar. These two scenes of moving down a ramp interpenetrate one another; they are part of an encompassing relationship. What Elton noticed is that they are, in some important sense, the same motion.

Asked to repeat what he noticed, Elton made a claim about the similarity of the two motions. He employed a markedly hypothetical language of "ifs" and "coulds": "If a car could get faster going down a ramp you could ..." He was not so much recounting his lived experience as seeing it in light of new possibilities: The pattern of his speed as he runs down a ramp could resemble that of a car rolling down a ramp. Again, the world of the rolling toy car and his running body interpenetrate, each illuminating the other in terms of his developing perspective on "change in speed" as the key feature of these two motions. Elton then asserted a warrant for his claim: "Because, um, um, when you get down and down it makes your legs go faster [revolves his hands in a rolling motion]." His gesture fused into one image the car wheels turning and his legs churning. Elton's juxtaposition of these two motions seems to have illuminated for him something about the experience of going "down and down." He explicitly linked the experience of the distance on the ramp being covered ("when you get down and down") with increasing speed ("it makes your legs go faster"). He had an embodied idea of the way this kind of motion unfolds and of what "go faster" might mean in a situation involving ongoing speed change.

Discussion

The word *everyday* often has the connotation of "common" or "unsystematic," as lacking analytic power for grappling with complex scientific phenomena. But in Elton's thinking, *everyday* experience proved highly generative. Elton engaged in *re-presenting* everyday experience—in seeing his experience of running down a ramp in new terms as he noticed a likeness between it and other instances of motion down a ramp. Wittgenstein (1953) called this kind of seeing (e.g., organized by contexts and patterns; the juxtaposition of prior histories with the here and now of interaction, beliefs, expectations) "seeing as." Wittgenstein meant to show us the multiplicity involved in seeing, the boundless possibilities of seeing one thing in terms of another, of exploring familiar terrain in new ways: "I contemplate a face, and then suddenly notice its likeness to another. I see that it has not changed; and yet I see it differently" (1953, p. 193).

Building on Wittgenstein, we would say that Elton did not simply recount what he already knew or experienced. Surely, in some sense he already knew that as he ran down a ramp he got faster. But here he seemed to know it in a new way, to be noticing new aspects, for example, that speed on a ramp can change in a certain kind of pattern. For Elton, this new way of seeing motion down a ramp involved inhabiting it from various perspectives: observed from outside a rolling toy car, imagined from inside a rolling car, recreated from his experience running down a ramp. Elton's account made the familiar strange, shedding light on his bodily experience as well as his observations of motion.

Through his embodied imagining, Elton came to see trips down a ramp as trips of a particular kind, in which cars and people got faster as they moved down the ramp. In this re-presenting, he linked the world of his everyday experience to the world of the toy car through a common relation:

the feeling of going faster down an incline. He rendered each in new terms as he noticed relationships between them in a way that was analytically generative for the central problem. He brought to light "family resemblances" (Wittgenstein, 1953) among trips in which any two share features in common. Through such juxtapositions and noticings, Elton and the rest of the class built up a sense of the "family" of trips down inclines, "a complicated network of similarities overlapping and crisscrossing: sometimes overall similarities, sometimes similarities of detail" (Wittgenstein, 1953, p. 32).

This practice of re-presenting everyday experience was prevalent throughout the class' investigation and strongly encouraged by Ms. Pothier, who seized the moment to elicit other children's experiences: the effort they needed to slow themselves while moving down a ramp, the difficulty of stopping near the bottom, and the pattern of increasing speed from slow (walking) to fast (jogging) to faster (running). The children's re-presentation created new possibilities for encountering and noticing different aspects of motion.

With Ms. Pothier's guidance, the children went on to investigate motion down a ramp using perceptual practices (Goodwin, 1994) developed in response to their emerging needs to make the changes in a toy car's motion visible. These came to be called "speed tests" and included such things as observing the gradual blurring of moving wheels on the toy car, listening for increasing loudness of sound, feeling and observing the effect of a collision between the car and a block (held at various positions along the ramp), and evaluating the car's speed as it passed through a series of tunnels positioned along the ramp. Like practicing scientists, Ms. Pothier and the children used their tests to structure the domain they were scrutinizing (i.e., motion down ramp) into a field of visible relevance (i.e., making a pattern of changing speed visible through varied and convergent means). Their work and increasing fluency with these practices supported development of increasingly comprehensive qualitative accounts of this motion as positively and continuously changing. At the end of this sequence, the children accounted for change in the car's speed in the following ways:

> We discussed that it was fastest at the end, because the speed builds up at each place.

> ... the more the car goes down the track, the more it gets faster.

> At the beginning it [the car] is just getting speed, in the middle it can still get speed because it has more time, but at the end there's no time left to get more speed.

> "I think it builds up at each place. It goes up from the beginning to the end.

In each of these cases, the students "see" the car's speed as continuously and positively changing, as "building up at each place." They are offering accounts that are grounded in complex acts of perceptions. The perceptual practices embodied in the "speed tests," along with the re-presentations of everyday experience in which the children engaged, created the ground for developing accounts that spanned all the instances of motion down a ramp that they invoked or encountered. The teacher and children's work to make change in speed visible connects with an overarching goal of Newtonian physics, namely, to develop universal explanations for diverse instances of motion. Although the children constructed an understanding of basic ideas of Newtonian mechanics including acceleration, force, and inertia, their work had a quality distinct from the way physics is typically taught: The world of their everyday experience and the world of physics were held mutually accountable. What they accomplished was to define "what it means to see" and "what there is to see" (Latour, 1986) in order to make the ordinary world of motion as they experienced it visible in new terms.

CASE 2: NEGOTIATING SHARED MEANINGS

A good deal of research has demonstrated the importance of engaging children in practices of theorizing, explanation, and argumentation in science (e.g., Brown & Campione, 1990, 1994; Herrenkohl, Palincsar, DeWater, & Kawasaki, 1999; Hudicourt-Barnes, 2003; Michaels & Sohmer, 2000; Rosebery, Warren, & Conant, 1992; van Zee & Minstrell, 1997; Warren & Rosebery, 1996). But, to our knowledge, little has focused on the kinds of difficulties in language and symbol use that we see emerging regularly as a part of these practices. Many of these difficulties result from contrast or conflict in what we might call, following Gee (1999), "situated meanings" among participants in the discussions crucial to scientific inquiry.

By "situated meaning," we are not referring to a definition, but to the practices that words and symbols regularly participate in and from which they gain the local meanings that individuals and communities use in particular contexts. A single word might be important in a variety of intellectual and social communities and in the activities those communities routinely engage in and, thus, have a range of situated meanings. The term *light*, for example, is part of different practices and concerns in astronomy and photography. Although as language users, we each have access to a variety of situated meanings for a given idea, we do not necessarily each have access to the same set of meanings. Children can have different views of which word to use in a context or what exactly a word means in a given context.

Conflicts arise when learners have a variety of meanings and contexts for particular words. Teachers often seek to mediate such disagreements by providing the appropriate situated meaning for the context, or by correcting children if they seem to be using the wrong meaning, rather than allowing the disagreement to continue and the children to negotiate a common understanding. In the following case, the value of disagreement in forging shared understandings is highlighted as students jointly interrogated the meanings of *speed* and *step* in light of each other's meanings and in relation to the mathematics of change. In so doing, they developed an analytical view of their everyday language and of the social and intellectual purposes that underlie the use of a particular set of terms in science and mathematics. They also deepened their understanding of, and ability to talk about, complex scientific concepts.

The Study

Mary DiSchino wanted to see what, if anything, her third and fourth graders could learn about the mathematics of change, in this case, relationships among distance, time, and speed. In constructing an investigation of distance, time, and speed, Ms. DiSchino drew on a number of resources, including "Patterns of Change," a unit in the *Investigations in Number, Data, and Space* curriculum (Tierney, Nemirovsky, Noble, & Clements, 1996). One of her goals was to enable her students to recognize the power of their own thinking and to understand that wrestling with confusion is central to learning (DiSchino, 1998). Each week, the class's activity and discussion was videotaped, and Ms. DiSchino, in consultation with researchers, would plan the following week's work in light of what the children had said and done. (See Rosebery, 2005, for an account of Ms. DiSchino's planning process.)

The Focal Episode

A few weeks into the unit, the children were taking "trips" down a 9-meter strip affixed to the floor of their classroom. The children initially included fanciful elements (e.g., "twirl three steps, stop to pick up a dog, then fall backwards") in their trips. Teams then developed trip representations (usually in a cartoon format) for other teams to reenact, usually with mixed success. Around the time of the focal episode, Ms. DiSchino started to constrain the kind of representations the children could use, introducing them to the idea of charts as representations of trips that could be read and followed by others. In the course of the first classroom conversation about a chart, the children explored their sense of the categories they needed. They eventually arrived at four: Kind of Step, Distance, Speed, and Pause. Then they proceeded to chart a trip they had written together: "Walk two steps. Run three steps. Powerwalk to the end." The children put *walk* under Kind of Step and 2 under Distance. For the second sentence, they put *run* under Speed and 3 under Distance. When they came to "powerwalk," one child suggested that *power* go under Speed and *walk* go under Kind of Step, but the consensus seemed to be to put *powerwalk* as a unit under Kind of Step (see Table 3.1).

When Sonja asked why *run* should be under Speed, a hubbub resulted. One child said, "*Run* isn't stepping." Another said, "It means speed." The first child then reasserted, "When you're running, you're not stepping," to which Sonja responded, "You *are* stepping." She elaborated her point: "Step, just the same as powerwalking, you're stepping." She seemed to be suggesting that for her the meaning of *step* applied to both *run* and *power-walk*. Karen pointed out that "in powerwalking, you can count your steps and how many you're taking, but in running you can't really do that." Karen was agreeing that *step* had the same meaning as *powerwalk* and *run*, but that there was a principled difference between *powerwalk* and *run* in this context: She could count the steps in one case and not in the other. Counting was important because distance was seen as the number of "runs" or "walks." The children discussed whether it was possible to count steps. Someone suggested that it was, but it would take a long time. Sonja then pointed out that they had written, "Run three steps," so they must be able to count steps.

Building on her knowledge of the children and the domain, Ms. DiSchino allowed the conversation to continue for two more class sessions. The children were pulled in one direction by their sense of these terms as mutually exclusive: *run* as a speed and *step* as a deliberate and countable way to move. At the same time, they were pulled in another direction by the design of the chart and the view of these terms implicit in it, that is, that all motions could be described in terms of both speed and kind of step. They uncovered many aspects of both speed and step as they discussed their concerns:

Bob: You can say I'm gonna walk to the store. You're not saying one mile an hour. And then you can say I'm gonna walk this certain speed.
Johan: Pretty much anything that moves has a speed.
Saintis: [moving his hands] And you can feel the air.
Gladys: You can stay still, but you're still breathing. That's a speed.

TABLE 3.1
Representation of a Trip Along a 9-Meter Strip in Ms. DiSchino's Classroom

Kind of step	Distance	Speed	Pause
Walk	2		
	3	Run	
Powerwalk	3		

Bob: The difference between running and walking, it's not only a speed. It's also a kind of step. You keep one foot on the ground when you walk.

The students' chart had a familiar structure. Many of us, both researchers and teachers, would have been strongly tempted to explain to the children how these terms should be used in the context of the chart. Yet, because Ms. DiSchino allowed them to explore their own situated meanings, the students were able to uncover and understand aspects of the conceptual underpinnings behind its structure. They came to see the multifaceted nature of common terms such as *run*. They made aspects of this meaning explicit in order to determine what was or was not relevant to the mathematics in question. Sonja pushed them to consider what needed to be compared and how the chart could help. Did *run* belong in the category of Step or Speed? Either was possible, but one made more sense in this context than another. Bob helped them to see some of the choices possible in language. Saintis and Gladys noted that they felt the air as they moved, thus adding to the experiences that could be brought to bear. In such conversations, participants analyze their everyday usage, enrich the connections with other ideas and experiences, and, thus, uncover aspects of the conceptual and practical basis for new usage.

As the unit unfolded, the children and Ms. DiSchino further developed and refined their chart. It gradually took on the formal features of a conventional distance, time, and speed table. Students came to distinguish between interval and cumulative distance, calculating both and comparing interval distances in order to get a measure of speed. They added a "speed change" column to show the change in speed that took place during a trip and came to use and interpret their charts with ease. Toward the end of the unit, these third and fourth graders made initial forays into creating and interpreting graphs of distance and change of speed over time. In end-of-the-year interviews, these students attested to the importance of their conversation. Some quoted their contributions to this conversation verbatim; others thought *through* the conversation as they made choices in completing new charts. (See Monk, 2005, and Wright, 2001, for further discussion of the children's activity and learning.)

Discussion

Goodwin (2000) provided insight into how a system of meaning integral to particular practices is learned. In his view, learning meaning depends crucially on participation in the practices in question. He noted, for example, that graduate students in archaeology must develop facility with the terms for color and texture of the soil in which artifacts are found as well as with the forms on which these characteristics are recorded. Although they make use of everyday words (e.g., *dark yellowish brown, sandy*), these forms cannot be understood or completed effectively by outsiders. They cannot be understood without participation in the practice in which they were developed, the "webs of socially organized, situated practices" particular to these archaeologists and their intentions with regard to their discipline (p. 22). Similarly, in order to use their chart, the children had to learn certain conventions that carried implications for what they saw and how their inquiry proceeded. In the context of the chart, Step was not only a deliberate movement, but also an analytic category applied to motion; it became a category within which both *run* and *walk* could fit. Likewise, Speed became something that both *walk* and *run* had. Within this classroom community, these terms developed a shared history of use and meaning.

Teachers like Ms. DiSchino who are able to take advantage of situations of contrast between familiar and less familiar language practices provide ways for their students not only to clarify terms but also to enter new conceptual territory. For Sonja and her classmates, learning was not simply a matter of connecting new meanings to old, but of interrogating situated meanings, familiar and unfamiliar, in light of one another, and in the process bringing to the surface assumptions of use, purpose, and context (Bakhtin, 1981; Wittgenstein, 1953). Bringing to the children's attention, as these discussions do, the ways in which language and symbolic practices more generally are used in a discipline is not separate from engagement with the knowledge of that discipline. Instead, it engages children in an analytical stance toward the ways of seeing that are characteristic of a discipline and toward its models of natural and physical phenomena, and thus serves as a powerful context for learning the central ideas of the discipline.

CASE 3: THE ROLE OF IMAGINATION IN EXPERIMENTAL DESIGN

Most studies of children's scientific reasoning have used constrained experimental tasks to assess the degree to which children's thinking conforms to the thinking of scientists, that is, to a canonical view of experimentation as a method for identifying and controlling variables, or as a syntax of rules and strategies for making valid inferences (Klahr, Fay, & Dunbar, 1993; Kuhn, Amsel, & O'Loughlin, 1988; Kuhn, Schauble, & Garcia-Mila, 1992; Schauble, 1996; Schauble & Glaser, 1990). Such tasks require that students identify causal and noncausal factors in a multivariate context. These tasks also tend to privilege logical inference, or hypothetico-deductive reasoning, as the ideal of scientific thinking. In our work, we view experimentation as an exploratory process of constructing meanings for emergent variables rather than as a process of logical inference through which students—or scientists—identify variables and uncover relationships designed into the experimental setup. As part of this third case study, we designed open-ended tasks and observed what children did and how they made sense of experimental situations. A different picture from that found in the psychological literature emerged of the sense-making resources children bring to the practice of experimenting and the ways they learn about experimentation as a knowledge-making practice.

Again we focus on embodied imagining, an aspect of children's sense-making not often included on lists of desirable scientific practices or habits of mind. Research in education (e.g., Gallas, 1995; Paley, 1987) and in the sociology of science (e.g., Keller, 1983; Ochs et al., 1996; Rheinberger, 1997; Root-Bernstein, 1989; Salk, 1983; Wolpert & Richards, 1997) shows that both young children and scientists readily engage in embodied imagining. Our research suggests that this familiar, everyday practice can powerfully support learners in understanding complex scientific phenomena (Rosebery, 2005; Warren et al., 2001), including experimentation and control of variables. Here we describe how one student's thinking helped expand our view of experimental reasoning, from a form of logic to a multifaceted practice in which one builds and imaginatively "inhabits" experimental worlds to explore, evaluate, argue, and revise knowledge claims.

The Study

The students in a fifth-grade Spanish–English bilingual class investigated the behavioral ecology of ants by configuring habitats for sustaining ants and discussing their inferences about the ants' behavior in each. Their studies were more observational than experimental but involved a variety of configurations and comparisons. As students explored, in their words, the ants' "preferences" (i.e., in terms of light and soil moisture), they observed ants and their behaviors, configured containers to explore how the ants might respond, and developed, justified, and challenged knowledge claims. We note that there was little explicit discussion in class of experimental practice in terms of variables, controls, and experimental inference. (For further analysis of students' learning, see Eggers-Piérola, 1996; Warren et al., 2001.)

The Focal Episode: Constructing Darkness

As the teacher, "Ms. Hernandez," interacted with the children, we noticed aspects of their thinking which struck us as outside conventional descriptions of scientific reasoning in experimental contexts. The children seemed to imagine themselves into the ants' world, as a way to argue knowledge claims against the constantly changing background of experimental designs. Our growing wariness of taken-for-granted distinctions between "everyday" and "scientific" ways of thinking led us to look closely at what the children were doing, and we became convinced that they were engaged in something significant, albeit unfamiliar to us, that was worth exploring in greater depth.

To do this, we interviewed students in groups of two or three. We asked them to design an experiment to explore ants' "preference" for darkness or light, a question discussed in class. For present purposes, we focus on the thinking of one student, Emilio, and the two students interviewed with him, Juan and Yolanda. Over the course of a 40-minute interview, the students generated several designs. Much of their work focused on configuring wrapped and unwrapped water tubes and soil within the space of an empty plastic container in order to create contrasting conditions of darkness and nondarkness.

All three of the students thought ants preferred the dark, as many students had maintained in class. Emilio held that ants were always in the dark: Given the choice of a wrapped tube or an unwrapped tube, ants consistently went to the wrapped tube. Juan supported the same claim differently: "When we put dirt in there, they—they were a little bit walking around, but almost all of them were under the dirt, in the darkness." He imagined the dirt through the ants' "experience," the darkness an ant experiences by being under the dirt. Yolanda agreed with Juan. When asked, Emilio agreed that ants "like" to go under the dirt because "it's dark under there."

The children were then asked to design an experiment to prove this assumption. In front of them, they had a large, empty plastic box. Following the model used in class, Emilio proposed connecting two boxes: one with dirt and a wrapped tube; the other with only an unwrapped tube (see Fig. 3.1, Design 1). Because the two-box design proved technically impossible to create, Emilio and Juan proposed new designs. Emilio essentially replicated his first design, placing the two tubes, one wrapped and the other unwrapped, in opposite corners of the box (Design 2a). Commenting on Emilio's design, Juan proposed a modification: They should put in "something like metal" to divide the box into halves (Design 2b). One half would include "a little bit of dirt ... with a water tube wrapped"; the other would include an unwrapped tube but no dirt. They could then see if the ants "try to go this way, to this side [gesturing toward the side with the covered tube and dirt]."

When asked how they would know "whether it's the dirt or ... the darkness, then?" Emilio responded immediately with a proposal to put a cover on the side with the wrapped tube and dirt ("so it could be dark"), in effect creating an unambiguously "dark side" in the box (Fig. 3.1, Design 3a). Juan agreed. When asked whether this new design should include dirt, Juan said yes, but Emilio disagreed, saying that "It should be without the dirt

(Design 3b) 'cuz maybe they'd like the dirt (only) then ... that's the reason why they're going over there." Emilio explained that ants might prefer dirt not for the darkness it affords, but "maybe to keep warm because they (go under it)."

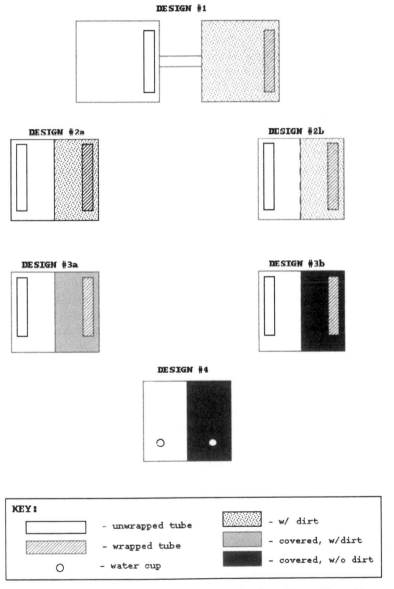

FIG. 3.1   The sequence of experimental designs developed by Emilio and Juan to test ant behavior in response to light and darkness.

Emilio positioned himself both inside and outside the ant habitat he was designing. He also envisioned, from the inside of this world, how its material elements might be experienced by the ants. Like Juan, who imagined ants "in the dirt, under the darkness," Emilio imagined how the ants might react to the presence of dirt and what that might mean in terms of their "preferences." He imagined the ants "going over there" into the dirt, "maybe to keep warm." The world he imagined was a lived-in world, one which he, the experimenter, entered imaginatively in order to experience it as an inhabitant: What might be so appealing about the dirt to an ant? What is dirt to an ant? How might an ant experience the dirt?

Emilio's embodied imagining from inside the ants' world was interwoven with the larger purpose of evaluating, from the outside, how ants might experience dirt in relation to a knowledge claim about their "preference" for darkness. His imagining was thus inextricably coupled to an evaluative stance he assumed toward his experiment, as a source of evidence for making knowledge claims. As he inhabited this world, he thought through the consequences of dirt for his claim about ants' preference for darkness and became newly concerned that the presence of dirt might interfere with any claims he could make regarding ants' preference for darkness. Dirt might possess a quality other than darkness, warmth maybe, that drew ants to the dirt. On this basis, he eliminated dirt from his design.

This interplay of inside and outside perspectives continued as Emilio considered other possible designs. The question of whether to include dirt persisted, with Yolanda wanting to add dirt to both sides of Emilio's design. Emilio addressed this issue: "If we put dirt, we should put dirt on both sides, but if—we *should not* put dirt (and) on no side(s), just put the thing in [using his hand to indicate a divider] and one side covered and the tube with the thing over it and a tube not with the thing over it." When asked why he would put dirt on both sides, Emilio explained that if they put dirt only on one side the ants might be drawn to that side by a quality of dirt other than its darkness. He then explained why he did not want to include dirt at all. If he did, then "both of the sides are still gonna be dark. One is just gonna be darker."

Emilio spoke from within different imagined worlds, evaluating each with respect to an argument about ants' preference for darkness. He worked with Yolanda's idea, knowing that he could not put dirt only on one side, but wondering if it would work to put it on both. At first Emilio spoke hypothetically: "If we put dirt, we should put dirt on both sides...." Conditioned perhaps by his earlier analysis of the problem, he implied that including dirt on both sides would control for its otherwise confounding effect. Then, in mid-utterance, he seemed to realize that the inclusion of dirt would not take care of the problem of restricting darkness to one half of the box. He quickly shifted out of the hypothetical space of Yolanda's de-

sign and negated it entirely: "We *should not* put dirt (and) on no sides ...."
He now imagined a space that was dirtless.

In Emilio's talk, imagined spaces and arguments about knowledge
claims seem indistinguishable; the boundaries between inside and outside
perspectives seem blurred. Emilio developed a feel for what ants might experience from the inside in relation to the components of the worlds designed and, from the outside, for his needs as a claimant in relation to these
worlds and to the question at hand. When he said that both sides would still
be dark—but one darker than the other—if they put dirt on both sides, he
was at once imagining these worlds in relation to ants' experience of them
and evaluating them in relation to the complication they posed for the
question under investigation.

During the interview, the students continued to work on this problem,
worrying about other features of their designs, in particular, the water tube
in Emilio's design. They came to see the water tubes as embodying, like dirt,
dimensions other than darkness: water, cotton, and a place to hide from potential threats. They worked out a way to provide ants with moisture, elimi-
nating the tubes and using instead a bottle cap with water, which Emilio
placed on both sides of the container (see Fig. 3.1, Design 4).

Discussion

The interplay of perspectives we have called inside and outside, or imaginative and evaluative, formed the ground of Emilio's reasoning. The inter-
twined perspectives meant he was "inhabiting" the world he was creating, in
order to envision events and outcomes, and simultaneously evaluating the
imagined events from outside this world, in light of possible knowledge
claims. This practice, which at another time in our history we might have
seen as anthropomorphism, was actually a generative practice for exploring the created world of the experiment in order to evaluate relevant knowledge claims. By imagining how ants might experience dirt, Emilio was able
in some sense to give them reason, which had consequences for the kinds of
evidence on which he was able to draw in making and supporting his claims.
Conversely, his emerging needs as a claimant gave shape to possible meanings of elements (or variables) in the world he was designing and imagining,
and the respective effects of these elements on the ants. Thus, Emilio's experimental activity involved a dynamic interplay of imaginative, evaluative,
material, and other resources, some of which are not traditionally considered central to the practice of experimentation.

This view is mirrored by findings from social studies and first-person
accounts of scientists at work, which reveal a similar heterogeneity of resources. The kind of embodied imagining analyzed here—of inhabiting a
created world to explore what might happen or how something might

react—is a feature of the everyday practice of scientists in both life and physical sciences (Keller, 1983; Ochs et al., 1996; Root-Bernstein, 1989; Salk, 1983; Wolpert & Richards, 1997). Nobel Prize winner Barbara McClintock, for example, described her state of being when doing some of her groundbreaking work in plant genetics:

> I found that the more I worked with them, the bigger and bigger [the chromosomes] got, and when I was really working with them I wasn't outside, I was down there. I was part of the system. I was right down there with them, and everything got big. I even was able to see the internal parts of the chromosomes—actually everything down there. It surprises me because I actually felt as if I was right down there and these were my friends … As you look at these things, they become part of you. And you forget yourself. (quoted in Keller, 1983, p. 165)

McClintock's powerful, imaginative mode of thinking enabled her, based on her extensive knowledge of maize, to visualize and describe a heretofore invisible biological process. Ochs et al. (1996), observing a similar practice among solid-state physicists, reported that it apparently allowed them to "symbolically participate in events from the perspective of the entities in worlds no physicist could otherwise experience" (p. 348). Although we are not suggesting that there is something isomorphic about embodied imagining across disciplines, studies of scientists at work suggest that imaginative participation in a represented world of biological, physical, or chemical events is a valuable aspect of scientific thinking, particularly when they are working through unresolved problems or exploring the implications of a model.

Emilio's interactive use of imaginative and evaluative thinking highlighted for us the value of this kind of thinking in children's scientific activity and expanded our view of children's scientific reasoning and our assumptions about the relationship between everyday and scientific practice. We do not see Emilio applying a form of logical reasoning to identify and control the variable "darkness," although the result of his work was a logically sound analysis resulting in a "controlled" experimental design. Instead, we see him constructing an increasingly more specified meaning of darkness by exploring how it related to the designed world of the experiment, the physical life of ants, and a discursive world of possible knowledge claims. Emilio created a stable, experimentally sound meaning for darkness out of potential meanings, a stability he achieved in interaction with the lived-in environment of the experimental world he was creating, in order, finally, to isolate it from dirt. We view this practice of imaginatively inhabiting the experimental world as a generative and powerful practice that both learners and scientists rely on in a variety of situations and for various purposes, whether it be to explore the created world of experiments to evaluate possible knowledge claims, to "get inside" important scientific ideas and models to see how they might work, or to develop and scrutinize theory.

STUDENT LEARNING AND ACHIEVEMENT

The three case studies presented here detail the deep understandings children developed through their classroom work. To complement these analyses, we wondered how children's learning in such classroom communities related to national standards and whether that learning would be reflected in measures of achievement. We addressed these questions as part of a larger design study involving the students in the first two case studies.

These children developed robust understanding of significant scientific ideas and practices typically taught to older students, exceeding the expectations set forth in state and national frameworks. The first and second graders (i.e., Elton and his classmates) learned about acceleration, force, and inertia—basic concepts of Newtonian mechanics that appear in the physics content standards for Grades 5 through 8 (Massachusetts Department of Education, 1995; National Research Council [NRC], 1996). The third and fourth graders (i.e., Sonja and her classmates) studied relationships among distance, time and speed, otherwise known as the mathematics of change, conceptual territory typically addressed in middle school (Massachusetts Department of Education, 1995; NRC, 1996; Tierney et al., 1996).

To assess student achievement, we administered four problems taken from a variety of high profile achievement tests and high quality curricula (the Massachusetts Comprehensive Assessment System [MCAS], the National Assessment of Educational Progress [NAEP], the Third International Mathematics and Science Study [TIMSS], Cambridge Physics Outlet [CPO]). The items were from the released versions of seventh- and eighth-grade levels of these tests and focused only on topics the students had studied in their classroom investigations (e.g., motion and force; distance, speed, and time; acceleration; gravity). Some items, originally written for seventh and eighth graders, were revised to be readable by the first and second graders.

The mean performance for the two items given to the first and second graders was 80% correct; the mean for the three items given to the third and fourth graders was 90% correct (see Table 3.2). Comparison data were available for two TIMSS items. Notably, both classrooms outperformed the international results for eighth graders by 28 percentage points for one item, and the third and fourth graders outperformed the international results for eighth graders by 9 points on another item. In considering these results, it is important to keep in mind that the students in these classrooms were from diverse backgrounds. Approximately half were from groups that

TABLE 3.2

Student Achievement Data From Assessment Composed of Released Test Items

Grade	Students (N)	Items given (N)	CPO Acceleration (8th grade)	TIMSS Gravity (7th/8th grade)	TIMSS Distance, time, speed (8th grade)	MCAS Gravity (8th grade)	Percent correct
1st–2nd	23	2	78%	83%			80%
3rd–4th	24	3		83%	92%	96%	90%
TIMSS International Results				49% (Gr 7) 55% (Gr 8)	78% (Gr 7) 83% (Gr 8)		

Note. CPO = Cambridge Physics Outlet; MCAS = Massachusetts Comprehensive Assessment System; TIMSS = Third International Mathematics and Science Study.

are historically underrepresented in the sciences and typically do not score well on science achievement tests.

CONCLUSION

To conclude, we return to the main theme of this chapter, the relationship between "everyday" and "scientific" knowledge and knowing. As noted in the introduction, a main tradition of research in the field of science education has viewed students' everyday experience and ideas as interfering with learning and needing to be replaced or repaired through instruction. This tradition casts the relationship between scientific and everyday as dichotomous. The scientific side of the contrast is typically associated with characteristics such as rationality, precision, formality, detachment, context independence, and objectivity. The everyday side of the term is characterized by a contrasting set of qualities: improvisation, indeterminateness, informality, engagement, context dependence, and subjectivity. The presumed differences between scientific and everyday are framed in opposition, with the left-hand term in the pair being the privileged one, the cognitive ideal: abstract versus concrete thinking, precise versus imprecise language, logical versus analogical reasoning, skepticism versus respect for authority, and so forth (Hymes, 1996). This contrast is broadly held in society, in education research, in schools, and even by some practicing scientists.

Hymes (1996) and Goody (1977), however, argued that although such binary formulations represent one way of bringing order into a complex world, "the order is illusory, the meaning superficial" (Goody, 1977, p. 36). Both of these scholars challenged the status of such dichotomies as descriptions of actual human experience. In a related move, Smith et al. (1993) critiqued the misconceptions tradition on a number of dimensions, the most important of these for our purposes being the overemphasis on the discontinuity between students' and experts' ideas. Rather than rejecting the relevance of prior learning from everyday experience in the physical world, Smith et al. (1993) sought to rethink students' conceptions as productive resources for conceptual growth, in line with basic tenets of constructivist theory.

In each of the case studies presented in this chapter, we see students taking a stance of inquiry toward what they know and have experienced as well as toward new aspects of meaning that emerge as they encounter ideas and practices central to science. They were encouraged along this path by teachers who designed their classroom instruction to emphasize practices of interrogating meanings of deceptively familiar experiences, such as motion down a ramp and ideas such as "faster and faster," "speed," or "darkness" in relation to situations of use and perspective. Out of these rich and, at times, conflicting encounters with meaning, the children revealed to themselves as well as to their teachers and to us the enormous generativity that connects what children know to what we want them to learn.

Conventional views of knowing value abstraction as a movement away or a detachment from the complex particulars of a situation, the extraction of a specific property or idea common to all instances of a given phenomenon. We are putting forward a different sense of knowing—one that emphasizes developing awareness of patterns of relationships fully embedded in the particulars of complex, lived and imagined situations (Arnheim, 1969). Rather than abstracting a concept like change in motion out of the many situations it is meant to describe, we see learners exploring such ideas by imaginatively inhabiting phenomena from different perspectives and actively interrogating meanings in relation to various contexts of use.

Wittgenstein (1953) described a concept as a thread, in which "the strength of the thread does not reside in the fact that some one fibre runs through its whole length, but in the overlapping of many fibres" (p. 32). We see learners creating such a thread, twisting fiber on fiber as they elaborate meanings of concepts in relation to scientifically relevant situations, imagining themselves into the objects or events they are trying to understand, and actively participating in the scenes they are creating. They take on both official and taken-for-granted everyday knowledge as something to be examined, interrogated, and populated with their own actions and intentions. In the cases presented here, children moved fluently among and through many "criss-crossing fibers," in the process making visible the generative potential of their everyday knowledge in the learning of complex ideas in science.

ACKNOWLEDGMENTS

This chapter is the result of our collaboration with colleagues—researchers and teachers—in the Chèche Konnen Center and would not have been possible without the ongoing conversation and joint work in which we all engaged. Particular thanks go to the teachers, Suzanne Pothier, Mary DiSchino, and "Ms. Hernandez," and to the students whose work forms the core of this chapter. We also wish to acknowledge the important contributions of Tracy Noble to the first case study and of Steve Monk to the second. Finally, we are grateful to our colleagues Faith Conant, Costanza Eggers-Piérola, Josiane Hudicourt-Barnes, Gillian Puttick, and Tracey Wright for their contributions to this work and to our thinking.

The design studies reported in the first and second cases were supported by the U.S. Department of Education, Office of Educational Research and Improvement, Cooperative Agreement No. R305A60007-98 to the National Center for Improving Student Learning and Achievement in Mathematics and Science, as administered by the Wisconsin Center for Education Research, University of Wisconsin–Madison. The design study reported in the third case was originally supported by a grant from the National Science Foundation (Grant No. RED 9153961). Subsequent analysis of the data and preliminary presentations of the third case were supported by the National Science Foundation (Grant No. ESI 9555712) and by the Spencer Foundation. The data presented, the statements made, and the views expressed are solely the responsibility of the authors. No endorsement by the foundations or agencies should be inferred.

REFERENCES

Arnheim, R. (1969). *Visual thinking*. Berkeley, CA: University of California Press.

Bakhtin, M. M. (1981). *The dialogic imagination: Four essays* (M. Holquist, Ed.; C. Emerson & M. Holquist, Trans.). Austin: University of Texas Press.

Ballenger, C. (1997). Social identities, moral narratives, scientific argumentation: Science talk in a bilingual classroom. *Language and Education, 11*(1), 1–14.

Ballenger, C. (2000). Bilingual in two senses. In Z. F. Beykont (Ed.), *Lifting every voice: Pedagogy and the politics of bilingualism* (pp. 95–112). Cambridge, MA: Harvard Education Publishing Group.

Brown, A. L., & Campione, J. C. (1990). Communities of learning and thinking, or a context by any other name. In D. Kuhn (Ed.), *Contributions to human development: Vol. 21. Developmental perspectives on teaching and learning thinking skills* (pp. 108–125). New York: Karger.

Brown, A. L., & Campione, J. C. (1994). Guided discovery in a community of learners. In K. McGilly (Ed.), *Classroom lessons: Integrating cognitive theory and classroom practice* (pp. 229–272). Cambridge, MA: MIT Press.

Clement, J. (1982). Students' preconceptions in introductory mechanics. *American Journal of Physics, 50,* 66–71.

Clement, J., Brown, D. E., & Zeitsman, A. (1989). Not all preconceptions are misconceptions: Finding "anchoring conceptions" for grounding instruction on students' "intuitions." *International Journal of Science Education, 11,* 555–565.

DiSchino, M. (1998). "Why do bees sting and why do they die afterward?" In A. Rosebery & B. Warren (Eds.), *Boats, balloons, and classroom video: Science teaching as inquiry* (pp. 109–133). Portsmouth, NH: Heinemann.

diSessa, A. (1993). Toward an epistemology of physics. *Cognition and Instruction, 10*(2/3), 105–225.

diSessa, A., Hammer, D., Sherin, B., & Kolpakowski, T. (1991). Inventing graphing: Meta-representational expertise in children. *Journal of Mathematical Behavior, 10,* 117–160.

Education Development Center. (1997). *Insights: An elementary hands-on inquiry science curriculum.* Dubuque, IA: Kendall-Hunt.

Eggers-Piérola, C. (1996). *"We haven't still explored that": Science learning in a bilingual classroom.* Doctoral dissertation, Graduate School of Education, Harvard University, Cambridge, MA.

Gallas, K. (1995). *Talking their way into science. Hearing children's questions and theories, responding with curricula.* New York: Teachers College Press.

Gee, J. P. (1999). *An introduction to discourse analysis: Theory and method.* New York: Routledge.

Gee, J. P., & Clinton, K. (2000). An African-American child's "science talk": Co-construction of meaning from the perspective of multiple discourses. In M. Gallego & S. Hollingsworth (Eds.), *What counts as literacy: Challenging the school standard* (pp. 118–135). New York: Teachers College Press.

Goodwin, C. (1994). Professional vision. *American Anthropologist, 96*(3), 606–633.

Goodwin, C. (2000). Practices of color classification. *Mind, Culture, and Activity, 7,* 19–36.

Goody, J. (1977). *The domestication of the savage mind.* New York: Cambridge University Press.

Hammer, D. (2000). Student resources for learning introductory physics. *American Journal of Physics: Physics Education Research Supplement, 68*(S1), S52–S59.

Herrenkohl, L. R., Palincsar, A. S., DeWater, L. S., & Kawasaki, K. (1999). Developing scientific communities in classrooms: A sociocognitive approach. *Journal of the Learning Sciences, 8,* 451–493.

Hudicourt-Barnes, J. (2003). The use of argumentation in Haitian Creole science classrooms. *Harvard Educational Review, 73*(1), 73–93.

Hymes, D. (1996). *Ethnography, linguistics, narrative inequality: Toward an understanding of voice.* Bristol, PA: Taylor & Francis.

Keller, E. F. (1983). *A feeling for the organism: The life and work of Barbara McClintock.* San Francisco: Freeman.

Klahr, D., Fay, A. L., & Dunbar, K. (1993). Heuristics for scientific experimentation: A developmental study. *Cognitive Psychology, 25,* 111–146.

Kuhn, D., Amsel, E., & O'Loughlin, M. (1988). *The development of scientific thinking skills.* San Diego, CA: Academic Press.

Kuhn, D., Schauble, L., & Garcia-Mila, M. (1992). Cross-domain development of scientific reasoning. *Cognition and Instruction, 9*(4), 285–327.

Latour, B. (1986). Visualization and cognition: Thinking with eyes and hands. In H. Kuclick (Ed.), *Knowledge and society: Vol. 6. Studies in the sociology of culture past and present* (pp. 1–40). Greenwich, CT: JAI Press.

Latour, B., & Woolgar, S. (1986). *Laboratory life: The construction of scientific facts* (2nd ed.). Princeton, NJ: Princeton University Press.

Lee, O., & Fradd, S. (1998, May). Science for all, including students from non-English-language backgrounds. *Educational Researcher,* 12–21.

Lee, O., Fradd, S., & Sutman, F. (1995). Science knowledge and cognitive strategy use among culturally and linguistically diverse students. *Journal of Research in Science Teaching, 32*(8), 797–816.

Lehrer, R., & Schauble, L. (2000). Modeling in mathematics and science. In R. Glaser (Ed.), *Advances in instructional psychology: Vol. 5. Educational design and cognitive science* (pp. 101–159). Mahwah, NJ: Lawrence Erlbaum Associates.

Massachusetts Department of Education. (1995). *Massachusetts curriculum frameworks for science and technology.* Boston, MA: Author.

McCloskey, M., Caramazza, A., & Green, B. (1980). Curvilinear motion in the absence of external forces: Naïve beliefs about the motion of objects. *Science, 210,* 1139–1141.

McDermott, L. C., Rosenquist, M. L., & van Zee, E. H. (1987). Student difficulties in connecting graphs in physics: Examples from kinematics. *American Journal of Physics, 55,* 503–513.

Michaels, S., & Sohmer, R. (2000). Narratives and inscriptions: Cultural tools, power and powerful sensemaking. In B. Cope & M. Kalantzis (Eds.), *Multiliteracies* (pp. 267–288). New York: Routledge.

Minstrell, J. (1989). Teaching science for understanding. In L. B. Resnick & L. E. Klopfer (Eds.) *1989 Yearbook of the Association for Supervision and Curriculum Development: Toward the thinking curriculum. Current cognitive research* (pp. 131–149). Alexandria, VA: Association for Supervision and Curriculum Development.

Monk, G. S. (2005). "Why would run be in speed?" Artifacts and situated actions in a curricular plan. In R. Nemirovsky, A. Rosebery, B. Warren, & J. Solomon (Eds.), *Everyday matters in mathematics and science: Studies of complex classroom events* (pp. 11–44). Mahwah, NJ: Lawrence Erlbaum Associates.

National Research Council. (1996). *National science education standards.* Washington, DC: National Academy Press.

Nemirovsky, R., Tierney, C., & Wright, T. (1998). Body motion and graphing. *Cognition and Instruction, 16,* 119–172.

Ochs, E., Gonzales, P., & Jacoby, S. (1996). "When I come down I'm in the domain state": Grammar and graphic representation in the interpretive activity of physicists. In E. Ochs, E. A. Schegloff, & S. A. Thompson (Eds.), *Interaction and grammar* (pp. 328–369). New York: Cambridge University Press.

Paley, V. G. (1987). *Wally's stories: Conversations in the kindergarten.* Cambridge, MA: Harvard University Press.

Rheinberger, H. (1997). *Toward a history of epistemic things: Synthesizing proteins in the test tube.* Stanford, CA: Stanford University Press.

Root-Bernstein, R. S. (1989). How scientists really think. *Perspectives in Biology and Medicine, 32,* 472–488.

Rosebery, A. (2005). "What are we going to do next?": A case study of lesson planning. In R. Nemirovsky, A. Rosebery, B. Warren, & J. Solomon (Eds.), *Everyday matters in mathematics and science: Studies of complex classroom events* (pp. 299–328). Mahwah, NJ: Lawrence Erlbaum Associates.

Rosebery, A., Warren, B., & Conant, F. (1992). Appropriating scientific discourse: Findings from language minority classrooms. *The Journal of the Learning Sciences, 2*, 61–94.

Salk, J. (1983). *Anatomy of reality: Merging of intuition and reason.* New York: Columbia University Press.

Schauble, L. (1996). The development of scientific reasoning in knowledge-rich contexts. *Developmental Psychology, 32*, 102–119.

Schauble, L., & Glaser, R. (1990). Scientific thinking in children and adults. In D. Kuhn (Ed.), *Contributions to human development: Vol. 21. Developmental perspectives on teaching and learning thinking skills* (pp. 9–26). New York: Karger.

Smith, J. P., diSessa, A. A., & Roschelle, J. (1993). Misconceptions reconceived: A constructivist analysis of knowledge in transition. *Journal of the Learning Sciences, 3*(2), 115–163.

Tierney, C., Nemirovsky, R., Noble, T., & Clements, D. (1996). *Investigations in number, data, and space: Patterns of change.* Columbus, OH: Scott, Foresman.

Ueno, N. (2000). Ecologies of inscription: Technologies of making the social organization of work and the mass production of machine parts visible in collaborative activity. *Mind, Culture, and Activity, 7*, 59–80.

van Zee, E., & Minstrell, J. (1997). Using questions to guide student thinking. *Journal of the Learning Sciences, 6*, 227–269.

Warren, B., Ballenger, C., Ogonowski, M., Rosebery, A. S., & Hudicourt-Barnes, J. (2001). Rethinking diversity in learning science: The logic of everyday languages. *Journal of Research in Science Teaching, 38*, 529–552.

Warren, B., & Rosebery, A. (1996). "This question is just too, too easy!": Perspectives from the classroom on accountability in science. In L. Schauble & R. Glaser (Eds.), *Innovations in learning: New environments for education* (pp. 97–125). Mahwah: Lawrence Erlbaum Associates.

Wittgenstein, L. (1953). *Philosophical investigations.* New York: Macmillan.

Wolpert, L., & Richards, A. (1997). *Passionate minds: The inner world of scientists.* Oxford, England: Oxford University Press.

Wright, T. (2001). Karen in motion: The role of physical enactment in developing an understanding of distance, time, and speed. *Journal of Mathematical Behavior, 20*, 145–162.

Chapter 4

Developing Algebraic Reasoning in the Elementary School

Thomas P. Carpenter
Linda Levi
Patricia Williams Berman
Margaret Pligge
University of Wisconsin–Madison

A growing consensus has emerged over the necessity to reconceptualize the nature of algebra and algebraic reasoning and to provide students opportunity to engage in algebraic reasoning earlier in their education (Kaput, 1998; National Council of Teachers of Mathematics [NCTM], 1997, 1998). The artificial separation of arithmetic and algebra traditional in school mathematics curricula deprives students of powerful schemes for thinking about mathematics in the early grades and makes it more difficult for them to learn algebra in the later grades (Kieran, 1992; Matz, 1982). The solution, however, is not simply to push the current high school algebra curriculum down into the elementary school. Rather, we need to broaden the conception of the nature of school algebra such that the emphasis is not merely on the learning of rules for manipulating symbols or the skilled use of algebra procedures, but on the development of students' algebraic thinking.

Algebra sometimes is portrayed as generalized arithmetic. This characterization provides only one perspective of the many facets of algebra (Kaput, 1998; NCTM, 1998, 2000), and it is not our goal to elevate this perspective over others. Our goal is to broaden children's conceptions of arithmetic, not to limit their perspective of algebra. We are proposing that the teaching and learning of arithmetic be conceived as part of the foundation for learning algebra, not that algebra be conceived only as an extension of arithmetic procedures. We are studying how to engage children in thinking about mathematics in ways that are more productive both for learning arithmetic and for smoothing the transition to learning algebra.

If students genuinely understand arithmetic at a level that they can explain and justify the properties that they are using as they carry out calculations, they have learned some critical foundations of algebra. Unfortunately, in the way that arithmetic is traditionally taught, many students develop the perception that arithmetic is simply a series of calculations, and they do not think much about the fundamental properties that make the calculations possible. By the time students begin studying algebra in high school, the basic properties of arithmetic are obvious to them, at least in the context of arithmetic calculations. But most students have narrow conceptions of the properties and how pervasive they are in arithmetic and algebra. Most high school students do not appreciate that a small number of fundamental properties underlie the calculations they have learned, and they do not see that the procedures they are using to solve equations and simplify expressions are based on this same small set of properties that they have used in arithmetic.

Our research shows that elementary school children can begin to engage in thinking about arithmetic in ways that emphasize and make explicit these fundamental ideas and practices that are the basis for understanding arithmetic and algebra. In other words, they can learn to think about arithmetic in ways that provide a better integration of the ideas of arithmetic and algebra. Our research focuses on two central themes that are at the core of this conception of what we, and others (e.g., Kaput, 1998), refer to as "algebraic reasoning": (a) making, justifying, and using generalizations; and (b) using symbols to represent mathematical ideas. Generalizing and representing generalizations involve the articulation and representation of unifying ideas that make explicit important mathematical relationships. These forms of thinking build directly on conceptions of understanding as constructing relationships and reflecting on and articulating those relationships (Carpenter & Lehrer, 1999). In fact they can be viewed as an attempt to further articulate how the development of understanding is instantiated in the elementary grades.

BACKGROUND

Previous research suggested that students in traditional classrooms are not aware of the underlying structure and properties of arithmetic operations (Chailkin & Lesgold, 1984; Collis, 1975; Kieran, 1989). Other researchers have found, however, that young children are capable of making such generalizations, constructing ways of representing them, and justifying them if they are provided appropriate opportunity (Bastable & Schifter, 1998; Davis, 1964; Kaput, 1998; Schifter, 1999; Tierney & Monk, 1998).

Schifter (1999) argued that algebraic thinking is implicit in the kind of meaning-making activities involved when students invent strategies for carrying out calculations. As children construct strategies for solving problems, they draw on fundamental properties of addition, subtraction, multiplication, and division and the relations among the operations. For example, some solutions for solving missing-addend (Join–Change Unknown) problems depend on understanding the inverse relation between addition and subtraction, and counting-on-from-larger strategies depend on commutativity of addition. It is a large step, however, to make explicit these fundamental understandings about number properties. Children who might readily use commutativity to solve problems with small numbers might express doubts whether the property holds for all numbers (Bastable & Schifter, 1998).

Teachers working with Shifter and Bastable described a number of cases in which generalizations about fundamental number properties arose spontaneously in their classes as students were solving problems. Often a strategy that a student used triggered a discussion of whether the strategy always worked and subsequently what that meant in terms of more general statements about a given operation. Building on students' spontaneous observations about regularities in number patterns and relationships in problem-solving strategies offers one context in which students might articulate generalizations about number properties, but early in our research we found that the opportunities did not occur with great frequency, and it was difficult to initiate them if they did not arise spontaneously. As a result, a major goal of our research has been to develop a coherent framework for constructing tasks that provide a context to focus on generalization.

STUDIES OF THE DEVELOPMENT OF ALGEBRAIC REASONING

For the last 5 years, we have worked intensively with a group of teachers to study the development of students' algebraic reasoning in the elementary grades and to construct instructional contexts that support that development. Drawing on Robert Davis' (1964) seminal work on the Madison Project, we have used the solution and discussion of true, false, and open number sentences as a primary context for helping teachers understand and focus on student thinking. These number sentences provide a context to initiate conversations that can lead to generalization and to introduce discussion of notation that might be used to express those generalizations. This context has proved particularly fruitful in engaging students in algebraic thinking and making that thinking visible to teachers. Specific ideas that we have addressed include (a) equality as a relation, (b) generalization about number and properties of operations, (c) representation of generalizations, and (d) the progression of forms of argument that students use to justify generalizations. The following discussion is based on work with 15 elementary school teachers and their students in Grades 1–6 (ages approximately 6–12), including design experiments in three classes.

Equality

Kieran (1992) characterized the distinction between arithmetic thinking and algebraic reasoning as a shift from a procedural perspective of operations and relations to a structural perspective. One of the hallmarks of this transition is a shift from a procedural view to a relational view of equality; developing a relational understanding of the meaning of the equal sign underlies the ability to make and represent generalizations. Behr, Erlwanger, and Nichols (1980); Erlwanger and Berlanger (1983); Kieran (1981); and Saenz-Ludlow and Walgamuth (1998) have documented, however, that children in the elementary grades generally consider that the equal sign means to carry out the calculation that precedes it; this is one of the major stumbling blocks when moving from arithmetic to algebra (Kieran, 1981; Matz, 1982). Our research demonstrated, however, that children can learn to think of equality in relational terms with appropriate instruction (Carpenter & Levi, 2000; Falkner, Levi, & Carpenter, 1999; see also Kieran, 1981; Saenz-Ludlow & Walgamuth, 1998).

At the beginning of the study, fewer than 10% of the students in any classes from Grades 1–6 demonstrated any evidence of relational understanding of the meaning of the equal sign (see Table 4.1). Given a problem like $8 + 4 = \square + 5$, the majority of students at every grade responded that 12 or 17 should go in the box to make the number sentence true. Teachers encouraged children to make explicit their conceptions and address the alternative conceptions held by different students in the classes. To provide a context for this discussion, teachers asked students to consider a range of true and false number sentences that challenged different conceptions of the meaning of the equal sign. For example, students would be asked to consider whether the following number sentences were true or false: $5 + 3 = 8$; $8 = 5 + 3$; $8 = 8$; $5 + 3 = 5 + 3$; $5 + 3 = 3 + 5$. Virtually all students agreed that the first number sentence was true. They were not so sure about the others, but as the unfamiliar forms of number sentences were introduced and compared to $5 + 3 = 8$, students were put in a position to examine and make explicit their implicit ideas about the ways the equal sign was used. In making their assumptions and conceptions explicit and discussing them with other students, some students began to question and change their conceptions of the meaning of the equal sign. The changes were not immediate

TABLE 4.1

Student Solutions to $8 + 4 = \Box + 5$, Before and After Instruction

Grade	Solution before instruction			Solution after instruction	
	12	17	7	12 & 17	7
	%			%	
1 and 2	58	13	5	8	66
3 and 4	49	25	9	10	72
5 and 6	76	21	2	2	84

Note. Percentages of students who solved the problem correctly are shaded.

or easy, but at the end of the year, the percentage of students in our study demonstrating a relational understanding of equality ranged from 66% in Grades 1 and 2, to 84% in Grade 6.

One of the striking features of these results is how little progress was made by students in traditional classes as they progressed through the grades. Sixth graders showed no more understanding of the meaning of the equal sign than did first graders. In contrast, with instruction that specifically challenged students to examine their conceptions of equality, students throughout the elementary school grades learned to use the equal sign to express relations rather than as a sign to carry out the preceding calculation.

Not all students who use the equal sign correctly, however, have the same degree of relational understanding that Kieran (1992) characterized as the hallmark of algebraic thinking. We identified two quite different ways that students thought about the relation expressed by the equal sign. These differences are illustrated by the responses of two students to problems with more than a single number after the equal sign. Both types of strategies are prevalent throughout the elementary grades as children acquire an understanding of the use of the equal sign as expressing a relation.

Kevin, a kindergarten student, solved the open number sentence $4 + 5 = \Box + 3$ by calculating the sums on each side of the equation. First he calculated $4 + 5$ by extending four fingers, then five more fingers, and counting the total. Next he figured out what number to add to 3 to get the same total. First he extended three fingers, then extended more fingers until he had a total of nine fingers. By keeping track of the number of fingers he added to the initial set of three fingers, he figured out that the answer was 6.

Donald, a first-grade student, used a strategy that went beyond simply calculating the sums on the two sides of the equation $6 + 2 = \Box + 3$:

Donald: It's 5.

Ms. F: How do you know it's 5, Donald?

Donald: It's 6 + 2 there. There's a 3 there. I couldn't decide between 5
                and 7. Three was one more than 2, and 5 was one less than 6.
                So it was 5.

To solve this problem, Donald did not actually carry out the calculation.
He recognized the relation between the numbers on the two sides of the
equation: 3 is one more than 2, so the number that replaces the box must be
one less than 6. Initially, he was not quite sure whether the number in the
box should be one more or one less than 6, but before responding, he
worked out that the increase in one number had to be offset by a decrease in
the other number.

Later in the period, Ms. F posed the following true-or-false number sentence: $57 + 38 = 56 + 39$. This problem offers more of a challenge in using
the kind of compensating strategy that Donald had used on the earlier
problem because the calculation is more difficult. Donald solved the problem by relating it to the problem he had solved earlier, pointing out the correspondences between the numbers in the two problems:

Donald: I know it's true because it's like the other one I did—6 plus 2 is
                the same as 5 plus 3.

Ms. F: It's the same as the other one. How is it the same?

Donald: 57 is right there, and 56 is there, and 6 is there, and 5 is there,
                and there is 38 there and 39 there.

Ms. F: I'm a little confused. You said the 57 is like the 6 and the 56 is
                like the 5. Why?

Donald: Because the 5 and the 56, they both are one number lower
                than the other number. The one by the higher number is the
                lowest, and the one by the lowest number up there would be
                more. So it's true.

The type of thinking that Donald exhibited in solving these two problems represents a significant advance over the thinking that Kevin used to
solve a similar problem. Both children viewed the equal sign as expressing a
relation between numbers. For both of them, the expressions on either side
of the equal sign represented the same number. But Kevin still viewed the
numbers on either side of the equal sign as representing separate calculations. He calculated the sum on the left side of the equation and found a
number to put in the box that, when added to 3, would give the same number. Donald, on the other hand, not only recognized that he was looking for
a relation between the sums on the two sides of the equation, but he also recognized a relation among the numbers in the two expressions that made it
unnecessary to actually carry out the calculations. Although both Kevin and

Donald demonstrated that they understood the appropriate use of the equal sign, Donald's strategy showed greater understanding of the relations between the expressions than Kevin's. Donald considered the relation between the two addition expressions in the equation, not just the relation between the answers to the two calculations. He was able to consider $6 + 2$ and $\Box + 3$ as more than calculations to be carried out. Furthermore, he not only recognized the relations among the numbers on the two sides of each equation, he also recognized how the two different equations were related.

Although the kind of relational thinking represented by Donald's strategy represents a significant conceptual advance over the calculation strategy used by Kevin, we have found that not all children initially use a strategy like Kevin's before they begin to engage in the type of relational thinking illustrated by Donald's strategy. In fact, we have observed some children use relational thinking to help them understand the proper use of the equal sign (Koehler, 2002). On the other hand, a number of children do initially use computational strategies like Kevin's before they are able to use more flexible relational thinking strategies. Although calculation strategies are sufficient to use the equal sign appropriately, thinking about relations is not just a way to simplify calculation. This ability to reflect on relations among mathematical expressions is critical for students to think more generally about arithmetic and to extend their knowledge of arithmetic to algebra. This kind of thinking provides the basis for generating and using generalizations.

Generalizations About Number and Properties of Operations

When students make generalizations about properties of numbers or operations, they make explicit their mathematical thinking. Generalizations provide a class with fundamental mathematical propositions for examination as well as an opportunity to open up students' thinking for analysis and discussion. Students have a great deal of implicit knowledge of properties of arithmetic operations, but they generally have not examined generalizations about properties of numbers and operations explicitly or thought systematically about them. The trick is to find problems that provide students a context to make their implicit knowledge explicit. Discussion of appropriately selected true and false number sentences provides such a context.

The following example is taken directly from a case study of a group of second-grade students (Carpenter & Levi, 2000). This example illustrates not only how the students made their implicit knowledge explicit, but also how they refined the language used to articulate the generalization. Students were asked whether $78 - 49 = 78$ was true or false:

Children: False! No, no false! No way!
Teacher: Why is that false?
Jenny: Because it is the same number as in the beginning, and you already took away some, so it would have to be lower than the number you started with.
Mike: Unless it was 78 – 0 = 78. That would be right.
Teacher: Is that true? Why is that true? We took something away.
Steve: But that something is, there is, like, nothing. Zero is nothing.
Teacher: Is that always going to work?
Lynn: If you want to start with a number and end with a number, and you do a number sentence, you should always put a zero. Since you wrote 78 – 49 = 78, you have to change a 49 to a zero to equal 78, because if you want the same answer as the first number and the last number, you have to make a zero in between.
Teacher: So, do you think that will always work with zero?

[Mike interprets the question as whether it was necessary to change the 49 to a zero.]

Mike: Oh, no. Unless you 78 minus, umm, 49, plus something.
Ellen: Plus 49?
Mike: Yeah. 49. 78 – 49 + 49 = 78.
Teacher: Wow. Do you all think that is true?

[All but one child answers yes.]

Jenny: I do, because you took the 49 away, and it's just like getting it back.

Essentially, the children generated another generalization ($a + b - b = a$), although they had not yet articulated it as a general rule. It is a somewhat more difficult generalization to articulate than the zero properties for addition and subtraction, so after some discussion of the specific example, the teacher returned to sums and differences involving zero, using the number sentence 789,564 – 0 = 789,564:

Children: That's true.
Teacher: How do you know that is true? Have you ever done that? Ann?
Ann: I will tell you. All those numbers, take away zero, you won't take away anything, so it would be the same number.

After another example in which the children immediately responded that 0 + 5,869 = 5,869 was true, the following discussion ensued:

Teacher:	So we kind of have a rule here, don't we? What's the rule?
Ann:	Anything with a zero can be the right answer.
Mike:	No. Because if it was $100 + 100$, that's 200.
Jenny:	That's not that we are talking about. It doesn't have just plain zero.
Ann:	I said, umm, if you have a zero in it, it can't be like 100, because you want just plain zero, like $0 + 7 = 7$.

After some additional discussion to clarify that the children were talking about the number zero, not zero in numbers like 20 or 500, the children were challenged to state a rule that they could share with the rest of the class:

Ellen:	When you put zero with one other number, just one zero with the other number, it equals the other number.
Steve:	Not true.
Teacher:	Wait. Let me make sure I got it. You said, "If you have a plain zero with another number." With another number? Like just sitting next to the number?
Ellen:	No, added with another number, or minus from another number, it equals that number.

The group collectively came up with the rule: "Zero added with another number equals that other number." They also came up with the generalizations "Zero subtracted from another number equals that number" and "Any number minus the same number equals zero." One student, Steve, came up with several generalizations about multiplication. The comments came up in response to Ellen's initial generalization about zero, which did not specify an operation:

Steve:	I wasn't thinking about the zero stuff plus another number equals that same number that we added to the zero, but, umm, I was thinking about if you were [inaudible] a number times zero would be zero.... $7 \times 0 = 0$. That is what I am trying to say, because 7 zeros or 0 sevens would be zero, and if you just add 7 zeros, you would just get zero.... Even a high number times zero would be zero.... Even the highest number you can think of times zero would be zero.
Teacher:	How do you know that, Steve?
Steve:	... Like 256 times zero, 256 zeros and any amount of zeros would be zero.

Steve had initially objected to Ellen's overly general statement of a zero rule. Having clarified the rules, we questioned Steve about whether his rule was an exception to the rules for addition and subtraction. He responded that those rules were like $27 \times 1 = 27$. This comment suggested that Steve had at least some level of understanding of the parallels between additive and multiplicative identities.

In these examples, children readily applied generalizations about zero to determine the truth value of number sentences. They not only applied them to solve given problems involving zeros, but they also came up with number sentences that embodied additional principles ($78 - 49 + 49 = 78$), and one student, Steve, spontaneously came up with generalizations that did not evolve out of solutions to specific number sentences. Children often tried to state their generalizations using a specific example (e.g., "It's, like, $7 + 0 = 7$"), and they often used specific simple cases to validate their generalizations. They all were confident, however, that their generalizations held for all numbers. The justification of their assertions generally referred to zero representing "nothing," but they did say that zero was a number. Ellen's initial statement of the zero principle was overly general and not accurate, but collectively the students identified the limitations and constructed more focused, valid generalizations. At this point, the generalizations were stated in natural language, which can be awkward and imprecise.

Another feature of the discussion was that the students seemed to demonstrate a good conception of the appropriate use of counterexamples to challenge other students' claims or generalizations. For example, Mike challenged Ann's generalization that "Anything with a zero can be the right answer" with the counterexample, "No. Because if it was $100 + 100$, that's 200." That forced Ann to revise her assertion to make it more precise. Steve also challenged Ellen's generalization about operations with zero by bringing in multiplication. The students had a little more difficulty articulating reasons that their assertions were true, but they generally seemed to recognize that a single case or several cases did not prove a statement was always true, and they attempted to use arguments that would apply to all numbers (e.g., "All those numbers take away zero, you won't take away anything, so it would be the same number," "Any amount of zeros would be zero").

True, false, and open number sentences provide a context for students to begin to make generalizations explicit. Although number sentences generated by the teacher provided the initial basis for drawing out generalizations, once the classes started to talk about generalizations, making generalizations became a class norm, and students proposed generalizations on their own. In most classes in the project, the class wrote the generalizations on sheets of paper and posted them in some location in the room. They often referred to the generalizations when they used them, and sometimes they returned to a generalization to edit it to make it clearer. Some generalizations were difficult to state clearly in natural language, and representing generalizations with symbols provided a way of expressing the generalizations precisely.

Representing Generalizations With Symbols

One of the outcomes of making the use of true, false, and open number sentences an integral feature of classroom activity is that it provides students access to a notation precisely for expressing generalizations using variables (Carpenter & Levi, 2000; Davis, 1964). Students who worked with open number sentences in flexible ways quite readily adapted them to represent generalizations. There were several ways that teachers encouraged students to do this. One of the ways that we have introduced the idea of using variables to express generalization has been to ask students to write an open number sentence that is true for every number they can put in (Carpenter & Levi, 2000). Students often do not immediately understand the task and generate number sentences that do not express generality. For example, with one group of students, one student initially suggested $1 + 1 = 2$, which was not an open number sentence. Another student suggested $\square + D = O$. He gave examples of numbers that could be substituted, resulting in true number sentences, but the teacher suggested other numbers that showed that the number sentence was true for some substitutions, but not for all. Another child suggested $\square + \square = \square$, but other children offered numbers that showed that this sentence was not true for all substitutions. After the students had tried and rejected a number of suggestions, the teacher provided some scaffolding by asking if there were any special numbers that might work. This question steered the students to begin writing generalizations involving zero such as $\square + 0 = \square$.

Following is an example of how representing generalizations symbolically emerged in one combined first- and second-grade class. The class was discussing whether "Zero plus a number equals that number" and "A number plus zero equals that number" were different conjectures:

Laura:	They are different. The first one is, like, zero plus number equals number [writes $0 + \# = \#$ on the chalkboard], and the other is, like, number plus zero equals number [writes $\# + 0 = \#$].
Ms. K:	Can anyone else write those conjectures with something other than a number sign? [Mitch writes "$* + 0 = *$" and "$0 + * = *$"; Carla writes "$\square + 0 = \square$" and "$0 + \square = \square$"; Jack writes "$m + 0 = m$" and "$0 + m = m$."]
Ms. K:	Jack just wrote $0 + m = m$. What were you using the m to mean?

Jack: I meant it to mean any number.
Ms. K: Then we could put any number in there? What if I put a 2 here
               and an 8 here?
Laura: No ... you could put a 2 here and a 2 here. You have to put the
               same thing in both places.

Following this discussion, the class revisited the generalizations they had posted around the room and wrote open number sentences for those that they could represent with open number sentences. Some generalizations, such as generalizations about adding even and odd numbers, could not readily be represented with open number sentences. This led some classes into a discussion of the differences between generalizations that the students could represent with open number sentences and those that they could not. This provided a context to distinguish between generalizations representing basic properties of number operations (field properties and theorems that could be derived from them) and other more restricted generalizations.

In individually administered interviews at the end of the year, 14 of the 17 students in the combined first- and second-grade class in the case study used symbols to express generalizations such as $t + 0 = t$ and $h - h = 0$. These findings were replicated in two case studies of fourth- and sixth-grade classrooms, in which 80% of the students used variables to express similar generalizations.

Not only did this use of variables provide students with precise and succinct notation for representing generalizations, but it also engaged them in using variables in ways that were more inclusive than was commonly the case throughout the curriculum. Kieran (1992) and Matz (1982) pointed out that most students' experience with algebra entails solving equations to find values for unknowns. This leads to very narrow conceptions of the ways symbols are used in equations and to some serious misconceptions about variables.

These misconceptions are illustrated by the difficulty that college students and other supposedly mathematical literate adults often have in solving the notorious Student–Professor problem (Clement, 1982; Kaput & Clement, 1979): "Write an equation using the variables S and P to represent the following statement: There are six times as many students as professors at this university. Use S for the number of students and P for the number of professors" (Kaput & Clement, p. 208). Only 63% of first-year college engineering majors and 43% of social science majors were able to correctly represent this relation ($6P = S$). The most common wrong solution was $6S = P$. This error turns out to be very robust and is attributed to fundamental misconceptions about variables (Clement, 1982; Rosnick & Clement, 1980). The same problem was administered to students in the fourth- and sixth-grade case studies. Eighty-four percent of the fourth-grade students and 74% of the sixth-grade students wrote the correct equation to express this relation ($6P = S$).

The Progression of Forms of Argument Students Use to Justify Generalizations

We have identified different forms of argument that children use to justify their conjectures as well as key tasks that elicit more advanced forms of argument. In the primary grades, children tend to justify propositions by example, but as they move into the third and fourth grades they begin to see the limits of arguing by example and begin to employ more sophisticated forms of argument. The following example, drawn from a year-long case study of a sixth-grade class (Valentine, Carpenter, & Pligge, in press), illustrates the progression of forms of argument that students use to justify commutativity of multiplication. The data reported here come from field notes and audio recordings of two class sessions and individual interviews administered in the spring.

In early November, the teacher, Ms. V, asked the students to determine what number could replace n to make the following number sentence true: $325 \times 6 = n \times 325$. Only one student, Daniel, explained the array model in his written response. Thirteen students wrote that the numbers must be the same because the equal sign means "the same as." These students explained in a variety of ways that both numbers should be represented on both sides of the equal sign. One proposed that, "If you take 325 from both sides, you only have 6 and n. So n equals 6." Three students explained that the products would be the same if n was 6. One student thought that n should be the product of 325×6, which suggested that she still interpreted the equal sign to mean "the answer comes next." Because of the large numbers involved, none of the students actually carried out the calculation.

In the class discussion that followed, most students provided ambiguous explanations about why $n = 6$. A number of students gave responses similar to Abby, who responded that "n equals 6 because you have one 325 on each side of the equal sign, and since both sides are equal, the other number on each side would be equal to each other." These students never mentioned switching the order of the numbers or the operation of multiplication.

For many of these students, the sameness of the two sides of the equation was the prominent feature of the number sentence. This is clearly illustrated in Karl's response: "I think it is 6 because the two 325s are the same, and so that means that the 6, the n has to be 6 if they are going to be the same thing." When asked directly whether the order of numbers could be switched for all operations, virtually all students did acknowledge that commutativity only applied to addition and multiplication, but at this point many of their explanations and justifications were not clearly focused.

Daniel, the one student who had provided a solid explanation for his response, explained that, with an array with 325 dots in 6 rows and an array with 6 dots in 325 columns, the same number of dots would be counted. Ms. V asked the class to look at 3 × 5 and 5 × 3 arrays. By the end of the session, about five students had demonstrated some understanding of Daniel's array model, but it was questionable whether they could apply it to larger numbers or more general cases. Furthermore, the argument was only applied to a specific case, and even Daniel had not publicly argued for a general proof of the commutative property of multiplication.

In early December, Ms. V gave assignments to small groups of students to write on one of three different tasks. Each group had a large sheet of grid paper to use to present their solution to the entire class. One group was asked to decide whether the statement 124 × 396 = 396 × 124 was true or false, to write a generalization for equations like this, and justify that the conjecture was true for all numbers.

The group wrote, "The conjecture for multiplication and addition is that it does not matter which order the numbers are in. The answer will always equal the same thing." This group then drew two arrays of 15 dots, one 3 × 5 and the other 5 × 3, and the discussion continued:

Jordan:	Here it is 3 × 5 [pointing to the array], and here it is 5 × 3.
Daniel:	It does prove that 3 × 5 equals 15, and so does 5 × 3 equal 15. But does it prove for every number you can do that?
Jordan:	Yes.
Arial:	It was just an example. It could have been 2 × 4, and it would have been 8.
Jordan:	I could give another example. See this [pointing to the dot arrangement] is 6. Here this is 3 × 5, and this is 5 × 3.
Ms. V:	Jordan, what do you think Daniel's question is?
Jordan:	How does this prove that it would work for any problem?
Ms. V:	OK, that's the problem.
Arial:	I can say now why it proved it. Because that's 5 and that's 3. So it shows that if you switch the numbers around, you would have the same answer, Even though you could have different, like, numbers like 4. What I mean by different numbers is not like 3 and 6 and 3 and 5. But I mean like 2 and 6 and 6 and 2, but, well, it's just showing that you can flip them around.
Daniel:	But then again, I can see why it works. I know it works. Actually, I am not sure why it works. I don't see why another one up there would help. I can see why. [Pauses] I just don't see any proof.

Arial:	There are a billion possibilities. Like 10 and 20 and 20 and 10 numbers. It is the picture that changes. Well, there would be more dots.
Karl:	With any number, you can just turn it. Like, if you make a dot array and you just turn it, it's *always* gonna have the same number of dots, no matter what.
Arial:	Yeah.
Daniel:	That's a little better. It's just that I didn't want more and more examples.
Abby:	Well, Daniel, you can spend a lot of time trying a lot of examples, and then you think you know it would work for any numbers.
Daniel:	I know. But then again there are an infinite amount of numbers, so it shouldn't matter. You need to show that it always works.
Ms. V:	Did you accept Karl's explanation?
Daniel:	I accepted Karl's explanation.
Jordan:	Take any group of dots [pauses], and flip it on its side. It will be the same number of dots.

Jordan and his partners used an array model to demonstrate that the order of numbers could be switched, but their model only illustrated a specific case. They did not take the further step of arguing that the process they had demonstrated could be applied to any number (at least any whole number). Daniel insisted on a new level of justification. He stated explicitly that numerous examples would not yield a proof. Jordan's suggestion that the array could be turned and not change the number of dots didn't satisfy Daniel. He probed until Karl explicitly stated that the process could be applied to any number. Jordan's final statement also acknowledged the necessity of making a general argument that applied beyond the specific case being illustrated. Arial, on the other hand, continued to talk about switching the numbers around, and it later became clear that she still had a superficial conception of commutativity.

In this episode, the students assumed almost complete responsibility for class discussion. Ms. V interjected only to clarify that the group had actually reached some consensus about making an argument about the generality of the justification. Although Ms. V did not guide the discussion, it remained focused on establishing norms for justification. Daniel played a key role in this discussion, actually assuming the role that one might expect of the teacher. He understood the limitations in the initial argument and pressed the other students to show that the argument they were using generalized to all whole numbers. Rather than simply stating the generalization himself, he asked a question of the other students to get them to articulate the principle.

In this class episode and the ones that precede it, we see how students negotiated norms for articulating and justifying generalizations. True, false, and open number sentences represented artifacts that provided a focus for discussion and a context that made it possible for students to assume substantial responsibility for deciding what counted as a legitimate statement of a generalization and what counted as justification. This sixth-grade class collectively established norms for justification that represented a great deal of progress over their norms for justification at the beginning of the year. At the beginning of the year, only one student, Daniel, demonstrated any indication that he understood the need for general forms of argument beyond specific examples. Although the class collectively generated valid justifications for generalizations, not all of the students could do so on their own. In individual interviews at the end of the year, 37% of the students demonstrated that they could generate a valid general justification of a generalization, but 79% identified examples of general justifications as preferable to individual number examples to justify that a generalization was true for all numbers.

CONCLUSIONS

The findings of our studies document that children throughout elementary school can learn to adapt their thinking about arithmetic so that it is more algebraic in nature. They can learn that the equal sign represents a relation, not a sign to carry out a calculation. They can learn to generalize and to express their generalizations accurately using natural language and symbols. Although not all students in the elementary grades will master mathematical proof, they can begin to engage in meaningful discussion about proof and make significant progress in understanding the nature and importance of proof.

The fact that children can learn to reason algebraically does not mean that they should, but we would argue that our studies provide insight into what is gained by taking advantage of young children's ability to learn to reason algebraicly and what is lost by failing to do so. If not addressed, misconceptions about equality and variable persist and provide serious impediments for learning high school algebra or other advanced mathematics (Matz, 1982). Understanding of justification and proof takes years to develop. In spite of the fact that many individual students in the sixth-grade case study were not yet able to generate proofs individually, most learned to recognize the limits of examples and the value of general arguments. They engaged in discussions of the nature of proof that made explicit important issues that most students never encounter at any point in their education. These experiences could provide a foundation for deepening their understanding of proof in the future.

Learning mathematics involves more than learning definitions, concepts, and skills. Being literate in any area entails understanding the forms of argument of the discipline. If students begin to engage in those forms of argument in the early grades, they can develop an entirely different identity for themselves in relation to mathematics. One of the things that was striking about the classes we worked in was that the students were engaged in sense-making. They thought that mathematics should make sense and that they could make sense of it. Students persisted for extended periods of time working on a problem because they thought they should be able to figure it out.

We note that all students benefit by engaging in the kinds of interactions that are required to make generalizations explicit, represent them accurately with natural language and symbols, and justify that they are valid for all numbers. Learning to use precise language and learning to communicate about mathematical ideas are important goals of the mathematics curriculum (NCTM, 2000). But there is an added benefit as well. The best students have always figured out generalizations, and by doing so they made mathematics easier to learn and apply. Making generalizations explicit so that they are available to all students can address important issues of equity and access to powerful ideas of mathematics.

REFERENCES

Bastable, V., & Schifter, D. (1998). Classroom stories: Examples of elementary students engaged in early algebra. In J. Kaput (Ed.), *Employing children's natural powers to build algebraic reasoning in the content of elementary mathematics*. Unpublished manuscript, National Center for Research in Mathematical Sciences Education, Madison, WI.

Behr, M., Erlwanger, S., & Nichols, E. (1980). How children view the equal sign. *Mathematics Teaching, 92,* 13–15.

Carpenter, T. P., & Lehrer, R. (1999). Teaching and learning mathematics with understanding. In E. Fennema & T. A. Romberg (Eds.), *Mathematics classrooms that promote understanding* (pp. 19–32). Mahwah, NJ: Lawrence Erlbaum Associates.

Carpenter, T. P., & Levi, L. (2000). *Developing conceptions of algebraic reasoning in the primary grades* (Research Report No. 00-2). Madison, WI: National Center for Improving Student Learning and Achievement in Mathematics and Science, Wisconsin Center for Education Research.

Chailkin, S., & Lesgold, S. (1984, April). *Pre-algebra students' knowledge of algebra tasks with arithmetic expressions.* Paper presented at the annual meeting of the American Educational Research Association, New Orleans, LA.

Clement, J. (1982). Algebra word problem solutions: Thought processes underlying a common misconception. *Journal for Research in Mathematics Education, 13,* 16–30.

Collis, K. F. (1975). *The development of formal reasoning.* Newcastle, Australia: University of Newcastle.

Davis, R. B. (1964). *Discovery in mathematics: A text for teachers.* Reading, MA: Addison-Wesley.

Erlwanger, S., & Berlanger, M. (1983). Interpretations of the equal sign among elementary school children. *Proceedings of the North American Chapter of the International Group for the Psychology of Mathematics Education*.

Falkner, K. P., Levi, L., & Carpenter, T. P. (1999). Children's understanding of equality: A foundation for algebra. *Teaching Children Mathematics, 6*, 231–236.

Kaput, J. (1998). Transforming algebra from an engine of inequity to an engine of mathematical power by "algebrafying" the K–12 curriculum. In National Council of Teachers of Mathematics (Ed.), *The nature and role of algebra in the K–14 curriculum* (pp. 25–26). Washington, DC: National Academy Press.

Kaput, J., & Clement, J. (1979). Letter to the editor. *Journal of Mathematical Behavior, 2*, 208.

Kieran, C. (1981). Concepts associated with the equality symbol. *Educational Studies in Mathematics, 12*, 317–326.

Kieran, C. (1989). The early learning of algebra: A structural perspective. In S. Wagner & C. Kieran (Eds.), *Research issues in the learning and teaching of algebra* (pp. 33–56). Reston, VA: National Council of Teachers of Mathematics.

Kieran, C. (1992). The learning and teaching of school algebra. In D. Grouws (Ed.), *Handbook of research on mathematics teaching and learning* (pp. 390–419). New York: Macmillan.

Koehler, J. (2002). *Algebraic reasoning in the elementary grades: Developing an understanding of the equal sign as a relational symbol*. Unpublished master's thesis, University of Wisconsin–Madison.

Matz, M. (1982). Towards a process model for school algebra errors. In D. Sleeman & J. S. Brown (Eds.), *Intelligent tutoring systems* (pp. 25–50). New York: Academic Press.

National Council of Teachers of Mathematics. (1997). Algebraic thinking [Special issue]. *Teaching Children Mathematics, 3*(6).

National Council of Teachers of Mathematics. (1998). *The nature and role of algebra in the K–14 curriculum*. Washington, DC: National Academy Press.

National Council of Teachers of Mathematics. (2000). *Principles and standards for school mathematics*. Reston, VA: Author.

Rosnick, P., & Clement, J. (1980). Learning without understanding: The effect of tutoring strategies on algebra misconceptions. *Journal of Mathematical Behavior, 3*, 3–27.

Saenz-Ludlow, A., & Walgamuth, C. (1998). Third graders' interpretations of equality and the equal symbol. *Educational Studies in Mathematics, 35*, 153–187.

Schifter, D. (1999). Reasoning about algebra: Early algebraic thinking in Grades K–6. In L. V. Stiff & F. R. Curcio (Eds.), *Developing mathematical reasoning in Grades K–12* (pp. 62–81). Reston, VA: National Council of Teachers of Mathematics.

Tierney, C., & Monk, S. (1998). Children reasoning about change over time. In J. Kaput (Ed.), *Employing children's natural powers to build algebraic reasoning in the content of elementary mathematics*. Unpublished manuscript, National Center for Research in Mathematical Sciences Education, Madison, WI.

Valentine, C., Carpenter, T. P., & Pligge, M. (in press). Developing concepts of justification and proof in a sixth-grade classroom. In R. Nemirovsky, A. Rosebery, B. Warren, & J. Solomon (Eds.), *The encounter of everyday and disciplinary experiences*. Mahwah, NJ: Lawrence Erlbaum Associates.

Chapter 5

A Teacher-Centered Approach to Algebrafying Elementary Mathematics

James J. Kaput
Maria L. Blanton
University of Massachusetts–Dartmouth

In significant measure, disappointing results on student achievement tests in the United States (National Center for Education Statistics, 1996, 1997, 1998) can be traced to our nation's approach to algebra, which is to introduce it abruptly and late, isolate it in courses separated from other mathematical subject matter, and teach it primarily as a series of procedural symbol-manipulation skills. Exacerbating this situation is the shortage of appropriately credentialed mathematics teachers, especially at the secondary level in urban districts (Glenn et al., 2000). The U.S. algebra problem has its roots in the separation of arithmetic from algebra, a separation that extends back to their separate origins and distinct historical purposes (Kline, 1972; Swetz, 1987). Although not new, this problem has become more acute and visible as algebra's role as filter and barrier to opportunity has been recognized (Moses, 1995) and as political forces have aligned to legislate algebra into every student's school experience in states across the country within one or another accountability regime. The great majority of the population's mathematics experience is now expected to include not only arithmetic, but also algebra.

The elements of a curriculum–content solution, which are rooted in the research of the 1970s and 1980s, can be found in an approach to algebraic reasoning expressed in national standards (National Council of Teachers of Mathematics [NCTM], 2000) and in related documents. They are the outcome of sustained critical and constructive analyses occurring over the previous dozen years (Bednarz, Kieran, & Lee, 1996; Kaput, 1998; Lacampagne, Blair, & Kaput, 1995; NCTM & Mathematical Sciences Education Board, 1998), including work by the National Center for Research in Mathematical Sciences Education (e.g., Kaput, 1999; Romberg & Kaput, 1999). These solution elements share the following features:

1. Algebraic reasoning is characterized by the dual abilities, on one hand, to generalize, justify, and express generality within structured symbolic forms, and on the other to use the structure of these symbolic forms to reveal deeper relationships and generalizations.
2. Algebraic reasoning involves many two-way connections with other mathematics, especially, but not exclusively, with arithmetic. Algebra is not merely a topic strand—it embodies an approach to all of mathematics that puts a premium on generalizing and formalizing.
3 Algebraic reasoning needs to develop over a long period of time in students' mathematical experience, beginning in the early grades and engaging most mathematical topics.

An approach to algebra embodying these features has the additional advantages of increasing proficiency with elementary mathematics, especially arithmetic (Kilpatrick, Swafford, & Findell, 2001), laying foundations for the increased mathematical learning needed at the secondary level and building the mathematical habits of mind that support that increased learning.

A SOLUTION STRATEGY: ENABLE ELEMENTARY MATHEMATICS TEACHERS TO ALGEBRAFY THEIR TEACHING

Our understanding of algebraic reasoning suggests that any real solution to the algebra problem must begin at the elementary level and be scalable and sustainable within the constraints (including accountability constraints) and capacities of teachers, administrators, and available instructional materials in the broad majority of U.S. school districts. The mathematical experience of most elementary teachers, however, is deeply oriented to teaching how to calculate correct answers to narrowly defined problems rather than to developing insight into mathematical situations (Thompson & Thompson, 1996). Teachers have typically had little experience with the generalizing and formalizing that we regard at the heart of algebraic reasoning. We are reminded of R. W. Hamming's famous dictum: "The purpose of computation is insight, not numbers" (1962, epigraph, p. 400). These shortcomings in mathematical experience are reflected in traditional curricula and materials, as well as in the requirements of typical preservice teacher education programs. Moreover, although standards-based materials reach beyond arithmetic to stress student understanding, they do not typically emphasize generalization and formalization, which we identify with algebraic reasoning. Instructional resources that support such skills, where they do exist, are typically at the "enrichment" margins, although some classroom resources have begun to appear in recent years (e.g., Ferrini-Mundy, Lappan, & Phillips, 1997; NCTM, 2001; Ruopp, Cuoco, Rasala, & Kelemanik, 1997).

For all these reasons, in our research we have focused on professional development for typical elementary school teachers within the constraints of their daily practices, resources, arithmetic orientation, and capacities to grow mathematically and pedagogically. In particular, we developed an explicit "algebrafying" strategy, which involves classroom-grounded teacher development that takes the form of cross-grade professional development seminars specifically designed to exploit teachers' existing material, curricular, pedagogical, and conceptual capacities in ways that build teacher knowledge in the following three dimensions:

1. *Instructional materials.* The building of algebraic-reasoning opportunities, especially those involving generalization and progressive formalization, from available instructional materials, including algebrafying common arithmetic activities and word problems by transforming one-numerical-answer arithmetic problems to opportunities for pattern- building, conjecturing, generalizing, and justifying mathematical facts and relationships within increasingly formal language and, thus, intentionally changing teachers' passive (if selective) relationship with instructional materials to one of active shaping and reshaping.

2. *Teacher understanding of student reasoning.* The building of teacher understanding of algebraic reasoning: building teachers' "algebra eyes and ears" so that they can identify and generate opportunities for generalization and systematic expression of that generality (including written expression) and then exploit these across mathematical topics.

3. *Classroom practice.* The creating of classroom practice and culture to support active student generalization and formalization within the context of purposeful conjecture and argument, so that algebraic reasoning opportunities occur frequently and are viable when they do occur. In this process, teachers (a) attempt to understand the particular forms of student thinking, representation, and argumentation central to building algebraic reasoning skill as well as the way those forms develop across grade levels; (b) work to enhance their instructional re-

source base while building their subject-matter knowledge; and (c) work to integrate and promote active understanding in the classroom and improvement of subject-matter learning for both themselves and their students.

Our goal was to build generative knowledge that could be sustained over the long term. The majority of the initial group of volunteers, with whom we worked for 2 years, eventually became seminar leaders. In what follows, we describe our professional development methods, clarifying what we mean by "algebrafying elementary mathematics"; offer a detailed view of what an "algebrafied elementary mathematics" looks like in elementary classrooms, the kinds of practices that embody it, and its impact on student achievement; and close by describing the impact of this professional development on our core set of teacher-leaders.

"ALGEBRAFYING" ELEMENTARY MATHEMATICS IN PROFESSIONAL DEVELOPMENT AND TASK DESIGN

As noted, most teachers lacked any direct experience with the kinds of algebraic-reasoning activity that we wanted them to be able to promote in their own classrooms. Our professional development, therefore, was intended to provide intense, authentic experience in algebraic reasoning that could be linked directly to their daily teaching and available instructional materials in ways that supported their systematic and collective reflection on the growth of these forms of reasoning across the grades.

We designed three kinds of professional development activities: (a) exploring and solving genuinely challenging mathematical problems for use in teachers' own classrooms, (b) customizing problems and sharing and reflecting on classroom trials of customized problems, and (c) algebrafying existing instructional materials, mainly from teachers' textbooks. Occasionally, teachers read and discussed short readings or viewed and discussed videos related to their ongoing work. The initial seminars for teacher-leaders led to a set of *Early Algebra Resources*, consisting of problems, teacher modifications of them, student work based on those problems, and teacher reflections on classroom experiences using the modified problems. These materials served as a renewable resource for the teacher-led seminars, during which participating teachers were explicitly expected to augment and refine the set of resources rather than merely "implement" them. In this section, we use the seminars as a backdrop to illustrate what we mean by "algebrafied elementary mathematics."

Exploring Algebraic-Reasoning Problems

Problems chosen for the seminars embodied important mathematical ideas, were approachable at different levels, could generate mathematically rich conversations, and typically involved substantial quantitative reasoning and computation. These problems provided a foundation of shared mathematical experience, a resource for classroom work and reflection, and experience in thinking about longitudinal growth of ideas and representational forms (Blanton & Kaput, in press). Some were well-used favorites that frequently occur in the literature, such as our algebrafied version of the handshake problem (Yarema, Adams, & Cagle, 2000):

> Twenty people are at a party. If each person is to shake everybody else's hand once, how many handshakes will be needed? How many handshakes will be needed for 21 people? How does the number of handshakes grow every time someone new is added to the group?

In this form, the problem has an easy entry via physical enactment and small numbers; is not readily solvable empirically (especially by elementary students), but begs for an analytical approach; and pulls (in the last of the problem) toward the kind of quantitative analysis that will produce a general solution. Seminar participants first worked in groups to solve the problems, then shared, compared, and contrasted different solution strategies. The seminar instructors paid special attention to the representational forms and arguments used, with a view to sensitizing teachers to potential variations in student thinking. Because these problems were quite challenging to most, they provided authentic experience of mathematical problem solving, contrasting with both the development of arithmetic procedures in the teachers' own classrooms and the development of student understanding of computational procedures commonly addressed in elementary mathematics professional development.

Expressing Generalizations Using Number Sentences. Another problem that illustrates concretely certain important features of our approach to algebraic reasoning involves the use of unexecuted number sentences. The following is an abbreviated account taken from a professional development seminar led by one of the authors based on a typical geometric *function–covariation* pattern problem taken in a somewhat different direction than is usually the case. The instructor drew on a white board a "4-triangle" (see Fig. 5.1a). Next to it, he drew a "5-triangle" (see Fig. 5.1b). He then asked the teachers to count the number of small "1-triangles" in each of the larger triangles. Although questions arose about the difference between "right-side-up triangles" and "upside-down triangles," the teachers had no difficulty in recognizing that there were 16 small 1-triangles in Fig. 5.1a and 25 in Fig. 5.1b. At this point, he asked the first generalization question: "So, if we had n little segments down a side, what we might call an 'n-triangle,' how many little 1-triangles would there be?" Teachers were quick to respond with "the square," "n-squared," "n times n," "the number squared."

In typical patterning activities, building a simple formula for the pattern generated, in this case, a simple quadratic function, ends the activity. Standard extensions of this problem consider how the numbers grow as rows are added to the triangle bottom, and so on, all in the service of developing the important function–covariation aspect of algebra (Kaput, 1999). The connection to arithmetic with which we were concerned, however, reaches beyond this to the *form* of arithmetic expressions: The activity is thus the *beginning* rather than the end of the exploration.

Beyond Patterning to the Algebraic Use of Arithmetic Symbols: Using Uncomputed Expressions.   Having referred to the darkened top of the triangle in Fig. 5.1c as a "2-triangle," the instructor asked, "How many 2-triangles are there in the 5-triangle?" As teachers explored this in groups, some again asked about upside-down triangles. The instructor suggested that they keep track of them separately. Two types of answers emerged: those counting only right-side-up triangles and those counting both kinds. The instructor then darkened the interior upside-down 2-triangle (see Fig. 5.2a). Consensus emerged that the two kinds of triangles should be counted separately, but that the total should include both. The instructor then asked the key question, which led to the next level of algebraic thinking: "How many 2-triangles are there in an n-triangle?" After about 10 minutes of noisy group work, most groups produced data tables, in which the independent variable was the size of the larger triangle and the dependent variable was either the total number of 2-triangles or the explicit two-term sum of right-side-up and upside-down 2-triangles. No group had a formula or a description of the pattern, but most had data for large triangles up to seven or eight units on a side; all noted that a triangle

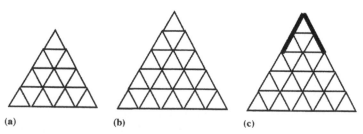

(a)   (b)   (c)

FIG. 5.1   A patterning activity using triangles.

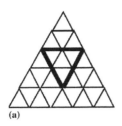 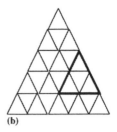

FIG. 5.2   Triangle activity. (a) A darkened "upside-down" 2-triangle inside a 5-triangle. (b) A darkened "right-side-up" 2-triangle inside a 5-triangle.

needed to have four units on a side in order to contain *any* upside-down 2-triangles. This complicated the patterns because zeros occurred for $n = 1, 2, 3$ in the upside-down accounting.

To build an appreciation for the way indicated sums can reveal underlying structure, the instructor led the counting of 2-triangles publicly for a group that had built a table out to $n = 8$, so that everyone could "see where your data come from." To make the counting as explicit as possible, he used a 3-finger "tripod" to identify a 2-triangle at the top of the configuration, with one finger at each vertex. He then slid this tripod down and to the left, then to the right, counting aloud. At that point, he said, "We need to count the next row, so how many will there be?" With his "tripod," he identified the darkened triangle in Fig. 5.2b, which he counted as "4," then slid to the left twice, counting and noting "we have six 2-triangles for the $n = 4$ big triangle." The explicit, physical, "counting-on" was the basis for the numerical sum.

At this point, Ms. A, an experienced third-grade teacher (who was adamant that she was "not good at math," yet frequently provided key mathematical insights) said, "The next row down will give you four more. If you take all of the right-side-up triangles, from top to bottom they increase." This provided the "teachable moment," an opportunity to make a first leap to an *indicated* sum. The instructor wrote "6 + 4," saying, "So we have 10. Is that right? Is that what everybody got for $n = 5$?" The groups agreed, and he filled out a hastily drawn "right-side-up" table for the five values of n (see Table 5.1), adding a sixth row and a column for the upside-down triangles. The instructor then asked, "How many upside-down 2-triangles in the $n = 5$ case?" He referred to the upside-down 2-triangle using his 3-finger tripod, saying, "Here's our first one, right? How many more?" The groups quickly agreed on two more, which he counted explicitly, then said, "So we have 3 upside-downs, which is what people were saying."

To bring out the growth substructure of the counting actions more explicitly, he then extended the existing figure to the $n = 6$ case: "How many

TABLE 5.1
Informal Counting Table

Up 2-triangles	Down 2-triangles
1	
3	
6	1
6 + 4	3
10 + 5	3 + 3

new right-side-ups will we get?" This refocused attention to the quantity being added on as n increased. The teachers, many of whom were drawing and counting, agreed there would be five more, and the instructor made a critical step: "So for $n = 6$, we have the old ones—10—plus the new ones—5." He then wrote "10 + 5" in the informal table he had drawn, deliberately writing the *unexecuted* sum.

Next, he asked how many new upside-down 2-triangles there would be, to which the teachers answered "3." The instructor noted that these were to be added to those already counted, which he recorded as "3 + 3" in the right-hand column of the $n = 6$ row. He then asked how many more of each type of 2-triangle would be added for $n = 7$. Response was mixed, with some giving totals and others giving the increments. He squeezed in an $n = 7$ row at the bottom of the table, with the executed totals, 21 and 10, which contrasted with the previous indicated sums.

He quickly asked what would be added on for $n = 8$. Making the point that figuring this out from *computed totals* would be harder than from *indicated sums*, he then erased the totals in the $n = 7$ row and replaced them with the 10 + 5 + 6 and 3 + 3 + 4. The answer was then evident from the *form* of the sums: 7 new up-triangles and 5 new down-triangles, which he squeezed in at the bottom of the table (10 + 5 + 6 + 7 and 3 + 3 + 4 + 5). The instructor then asked if teachers could make the table more consistent with the earlier entries. This led to Table 5.2, in which the earlier counting was repeated for $n = 1, 2, 3,$ and 4.

Extending Prior Understanding Using Unexecuted Sums and the Shapes of Formal Symbols. After discussion of the ways the counting actions were reflected in the sums, the information hidden by *computing* the sums, and the fact that, by using parentheses and labels, the part of a sum associated with a given triangle could actually be identified, the instructor posed a direct extension ("How many 3-triangles are there in a big triangle, one with

TABLE 5.2

Formal Table for *N*-Triangles, Using Indicated Sums

Number of up 2-triangles	*Number of down 2-triangles*
0	0
1	0
1 + 2	0
1 + 2 + 3	1
1 + 2 + 3 + 4	1 + 2
1 + 2 + 3 + 4 + 5	1 + 2 + 3
1 + 2 + 3 + 4 + 5 + 6	1 + 2 + 3 + 4
1 + 2 + 3 + 4 + 5 + 6 + 7	1 + 2 + 3 + 4 + 5

10-unit segments on a side?") to reinforce the value of indicated sums in revealing structure and relationships—in other words, *reasoning with the shapes of formal symbols*. They attacked the problem with great excitement, sensing that they had something new at their disposal—the understanding that the sequence of indicated sums actually revealed the pattern. Few teachers completed the problem in the little time remaining, but they were confident they could work it out. They had experienced directly the power of an algebraic use of arithmetic symbols: that the *form* of the symbols could reveal structure and deepen insight into a situation in a way that numerical computation did not. We ended with a brief discussion of ways to use the problem in their classes; the next session began with reflections on the use of unexecuted sums and number sentences.

Summary: Reflections on the Use of Unexecuted Sums.   Perhaps the greater power of mathematical symbol systems (and one strength of algebra over arithmetic) lies in the ways that systems express structures and forms, including the forms of computations. Thompson (1993), in the context of students' attempts to compare magnitudes of differences, pointed out that students' propensity to execute difference statements prevented them from reasoning with differences as conceptual objects. In the previously described vignette, we developed symbolic forms in the context of trying to develop insight into a geometric situation and to establish the validity of assertions. Our deliberate (and profitable) use of arithmetic expressions grounded in physical acts of counting and computing, but *not* actually computing, *extended the teachers' understanding of arithmetic*. This suspension of computation and attention in deference to the *forms* of the expressions was new in their thinking. By attending to the forms of the sums as inscriptions rather than as instructions to carry out procedures (see Thompson & Thompson, 1987), teachers were beginning exactly the kinds of structural analyses that most students do not experience, either in arithmetic or in algebra. This dearth of experience with the forms of symbolic expressions leads to a wide variety of well-documented error patterns (Lewis, 1981; Lins, 1992; Sleeman, 1984; Thompson & Thompson, 1987; Wagner, 1981; Wenger, 1987). We used activities like the ones described to provide teachers authentic mathematics learning experiences (lacking in most elementary teachers' backgrounds) in a way that links deeply to their arithmetic agenda.

Using Algebrafied Materials in Class

In the second core activity, teachers were asked to tailor the seminar problems and the algebrafied arithmetic materials to the grade levels they taught, to try their versions with their own students, and to bring in their own and their students' work on the problems, as well as written reflections on what happened in the classroom. The decisions teachers made modifying the problem brought their mathematical experience into direct contact with their teaching and their knowledge of students' thinking and learning. The resulting discussions inevitably made visible issues regarding teacher expectations, classroom norms, and content-specific knowledge. For example, most teachers focused strictly on counting the numbers of 1-triangles in larger triangles. The kindergarten and first-grade teachers made cut-outs of the small triangles to "cover" the larger triangles and focused on counting the number of little triangles in each row, putting the total alongside. Grade 2 teachers wrote out the sum for each large triangle. Grades 3 and 4 teachers found the totals as well as rules for the totals. Ms. A, for example, asked her students to determine the size of a triangle that had 36 little triangles in it. She also asked students how many little triangles the next bigger triangle would have and focused on showing that the bottom row could be written as a sum (36 + 13), which eliminated the need to count all the "old" triangles all over again.

Because we deliberately included teachers from multiple grade levels in the seminars, teachers were led to cross-grade comparisons, which helped them identify growth trajectories, general strategies for scaling problems up or down in difficulty, and ways to "bend" problems to serve important topics in their curricula. Teachers were encouraged to identify representational strategies (e.g., make aspects pictorially concrete, use physical manipulatives) useful in creating versions that younger students could attack productively. Teachers repeatedly were surprised with the performance of their students, especially their willingness to repeat drawings, counting, and computations for ever larger numbers. Eventually, the teachers also decided which shared materials (including teacher reflections) to add to the teacher resource materials.

Algebrafying Existing Problems: Transforming Arithmetic to Algebra in Flexible Ways

The third core activity in the seminars involved teachers algebrafying problems from their instructional materials, especially those from their basal textbook series. In particular, teachers were explicitly engaged in transforming frequently occurring arithmetic problems in two ways: (a) varying one of the numerical "givens" of the problem and (b) helping students write unexecuted number sentences that described the relationship or computation of the problem. The solution activity consisted of writing sequences of number sentences and examining the patterns in those sequences.

To illustrate, we use the following one-answer, deliberately mundane, arithmetic word problem: "I want to buy a T-shirt that costs $14 and have $8 saved already. How much more money do I need to earn to buy the shirt?" The first shift is in task definition: requesting a number sentence to express the relationship among the quantities of the given situation, a small shift with major consequences. Most typical solutions look something like $14 - 8 =$ money needed, or $14 = 8 +$ money needed, or $8 +$ money needed $= 14$. Accompanying this is a shift in the *use* of the number sentences produced from primarily supporting a computation that "closes" the problem to serving as the basis for variations on the problem situation. Determining a correct numerical solution to the initial problem is now the beginning rather than the end of the task. The core of the algebrafied problem is to use the number sentence(s) as the basis for the second major shift in the task definition: *treating the task as an occasion for building and expressing generalizations about the initial situation*—in any of several directions. For example, we could allow the *price* of the item to vary, and request a number sentence for each value: "Suppose it cost $15, or $16, or $17, or $26."

After several number sentences are written, the need for efficiency, coupled with the teacher's scaffolding, pulls toward the use of an abbreviation for the price of the T-shirt. The teacher can then ask what number sentence would show how much money needs to be earned for *any* T-shirt, regardless of price. The combination of these task design and pedagogical factors constitutes the algebrafication of the original problem, which deeply and subtly transforms arithmetic reasoning into algebraic reasoning.

Importantly, the algebrafication approach is also very flexible and can be extended:

> Assuming I make $2 per hour for babysitting, how many hours do I need to work to have enough money to buy the $14 shirt? If it cost $20? If it cost $P? Or, if I earned $3 per hour, how many hours do I need to work to buy the $14 shirt?

Most problems can generalize in many ways, and this source of flexibility enables a teacher to apply the algebrafying strategy to serve a wide range of curricular and pedagogical objectives and student needs. For example, the instructional goal might simply be to provide a context for skip-counting, or to introduce division, or practice subtraction, or to introduce letters as an efficient, compact way to write many statements at once. Or the goal might be more ambitious, for example, to introduce syntactical equivalence by generating multiple sequences of number sentences that express the same relationship. Importantly, algebrafying *brings to elementary mathematics an intrinsic feature of mathematics itself that accounts for mathematics' extraordinary power.* Because this work is grounded in teachers' everyday materials, teachers are pulled (a) to build from their existing instructional resource base and (b) to relate the approach to their daily practice rather than treating such tasks as "enrichment."

Adapting Problems for Different Grade Levels. One component of the seminars involved teachers' adapting problems to their respective grade levels and using the subsequent implementation as a focus for seminar discussions. This process enabled teachers not only to see the forms of mathematical thinking that occurred across the grade levels, but also to see the mathematical connections that *could* occur across grades and their *own* role in building these connections. To illustrate, we discuss here a first-grade teacher's adaptation of the outfit problem (also coincidentally a released fourth-grade test item in the data and probability strand from the 1999 Massachusetts Comprehensive Assessment System [MCAS]):

> What is the GREATEST number of outfits you can make with 2 pair of pants and 5 shirts? (Assume an outfit has exactly one pair of pants and one shirt.)

In preparation, "Gina" created colored cutouts of shirts and pants—tactile manipulatives students affixed to the easel on which she recorded their comments. In her reflections, she noted:

> I asked, "How many outfits can you make with this one red shirt, red pants, and blue pants?" Many children responded, "One." At this point, I needed to identify the meaning of *outfit*. Most wanted to put like colors together. Each pair of partners then was instructed to pair red with red, and then blue with red. "Oh! Now I know." Then the action began. We then recorded the information onto a chart paper. Another shirt was added to the two pairs of pants. Partners shared information and again recorded onto the chart paper. I recognized that they really needed to see this visually. Children began making predictions that one more shirt would be three more outfits. Matthew and Nathan, who are very confident with

their math skills, began shouting out their predictions without manipulating the outfits. Nathan said, "If you add one more shirt, you get three more because three plus three is six!" Matthew chimes in, "I know that!"

After several activities and adding of pants and shirts, the children were anxious to make predictions. We then used three shirts and four pants. Nathan predicted 14. Cory predicted 10 and then said, "Let's do it to see." Kyle, "I wish I had all these outfits." Matthew, "Well, if one shirt and three pants is three outfits, then we just have to keep adding three." ... After completing up to three shirts, the children began seeing a pattern in the third column of outfits. Many responded, "We are counting by twos." Nathan recognized that 2 + 2 is a double and wondered if three shirts and three pants would be six.... He was able to figure out there were nine outfits. We revisited the graph, and then some saw the increase by one in the first column. I wanted to get Nathan back to the doubles. We went back to one shirt and two pants and one doubled is 2. Two doubled is 4. Three doubled is 6. Then more children began seeing the pattern emerge.

Enacting the task through the use of paper outfits and organizing the data enabled students to build generalizations about patterns. For example, when varying the number of shirts but keeping the number of pants fixed at 2, students generalized that the number of outfits was increasing by 2s, the number of pants remained constant, and the number of shirts increased by 1s. Some students were eventually able to predict that 5 shirts and 2 pairs of pants would result in 10 outfits. We find these *kindergarten* students' performance to be significant given that the problem was a fourth-grade MCAS assessment item.

This example not only illustrates this particular teacher's capacity to render a task developmentally and mathematically appropriate for her young students, but it also indicates that algebrafying can extend to virtually any topic and that seminar tasks can live at various levels in elementary school mathematics. We also note that teachers in the seminars adapted problems on a regular basis and that the process became routine for most of them by their second year.

Summary: Design Principles of Tasks That Support Algebraic Reasoning. Four task design principles emerged collectively from teachers' and our own contributions to the seminars:

1. Provide easy entry via enactive counting or simple measurement through a sequence of events or objects that are definable, countable, or measurable.
2. Use repeated sequences of computations that yield numerical patterns that engage students arithmetically.

3. Promote unexecuted numerical expressions and number sentences that lie behind the patterns to serve as formal objects for algebraic reasoning.
4. Facilitate the algebraic use of number and operations on numbers beyond students' current computational range.

Addressing Classroom Practice and Norms

Over the course of the seminar, differences in teachers' ability to engage students in conjecturing and purposeful argumentation emerged. To address this, we repeatedly asked teachers whether a culture of inquiry was developing, what the classroom norms for argumentation were, whether students questioned each other and expected justification of mathematical statements, if students were learning oral and written communication skills, whether there were differences across classrooms, and how the teachers felt about the evolution of their practice. Most teachers passed through a transition period, treating their trials of modified seminar problems as separate from their "real" teaching or using these as enrichment or pullout work for special students. Most gradually expanded their repertoire of algebrafying skills to embody increasing numbers of the design principles, which they applied to increasing numbers of topics. A minority, whose practice was more rigid and computationally oriented, found it difficult to move beyond using pre-crafted tasks as pullout activities.

FEATURES OF ALGEBRAFIED ELEMENTARY TEACHER PRACTICE AND THEIR IMPACT ON STUDENT ACHIEVEMENT

In this section, we provide glimpses of a third-grade classroom taught by a teacher who was among the earliest to assimilate the algebrification approach. As a particularly strong example, her practice includes many of the features by which we measured progress toward an algebrafied elementary mathematics practice. We previously described (Blanton & Kaput, 1994) the degrees to which core teacher-leaders' practice embodied algebrafied features and how these changes reflected larger changes in their capacity for continuing growth and their relationship with their instructional resources. A much fuller account can be found in Blanton and Kaput (in press). In order to illustrate the impact on student achievement of this kind of strongly algebrafied practice, we compare achievement data on students in this teacher's class to data on students in a traditionally taught third grade in the same school, using 14 test items selected from the released items from the two previous years of the fourth-grade MCAS. We also compare the data on the "algebrafied" students with the data on students across the district and state.

A Case Study of Algebrafied Elementary Mathematics Practice

We observed and recorded "Jan's" 90-minute third-grade mathematics class approximately twice a week for the first full year of her participation in our professional development. The data (classroom field notes, audio recordings, Jan's reflections, students' written work, and classroom activities) were collected from 38 classroom visits. Nineteen additional classes were documented by Jan's written descriptions and reflections as well as by students' work. The socioeconomic status (SES) of this class and the school was lower than average for the district, with 75% on free and 15% on reduced lunch, 65% with parents for whom English was a second language, and 25% with no parent living at home. The 204 episodes of algebraic reasoning (observed in 57 classes) were separated into planned (35%) and spontaneous (65%) episodes (from 2–3 minutes to 30 minutes or more in length). About a third of the longer episodes were derived from seminar activities; the majority of the rest consisted of algebrafied arithmetic problems from the text or extracted from resources Jan found. We saw no pattern in the distribution of planned versus spontaneous episodes.

Our study of Jan's classroom also enabled us (a) to develop exemplar classroom episodes to feed into the seminars when teachers reported on related classroom activities and shared instructional materials, student artifacts, and written reports with the larger group; (b) to examine how ideas developed and promoted in the seminars played out in an actual classroom, yielding a micropicture of algebrafied classroom practice and teacher knowledge; and (c) to obtain a growing corpus of material, including video, for the seminar resource base. The low SES of these students provided additional credibility to their and their teacher's robust performance, credibility that fed the confidence of the other participating teachers and their principals.

Categories of Algebraic Reasoning

We coded the episodes into three broad categories, reflecting our overall analysis of algebraic reasoning (Kaput & Blanton, 2001), but these are not mutually exclusive and often occur in combination.

Algebraic Reasoning as Generalizing Arithmetic.   This category includes *exploring and generalizing about properties and relationships of whole numbers,* such as properties of odd and even numbers, the computational features of the base-10 placeholder number system, and the *algebraic use of numbers.* The discussion of a homework problem on the addition of whole numbers illustrates frequently occurring phenomena in Jan's class:

Jan:	Suppose you had 5 + 7. Would your answer be odd or even?
Student:	Even.
Jan:	How did you get that?
Student:	I added 5 and 7 and then I looked over there [indicating a chart on the wall containing even and odd numbers] and saw that it was even.
Jan:	What about 45678 + 85631? Odd or even?

Jan chose numbers sufficiently large so that students could not rely on computation to determine parity, had to examine structural features to reason whether a sum would be even or odd, and were led to focus on the properties of evenness and oddness and (implicitly) to treat the numbers as abstract placeholders. Jan and her students frequently had natural, spontaneous algebraic conversations based in arithmetic. Over the year, she returned to properties of odd and even numbers several times, asking, for example, "What can we say about the sum of three odd numbers?" The resulting discussion in this case built on previously established results: The sum of two odds is even, and the sum of an even and an odd is odd. In revisiting certain themes such as properties of odd and even numbers, Jan built both on her students' understanding of the properties of numbers and on their ability to generalize and justify their generalizations. Addressing algebraic reasoning in the context of elementary mathematics, as Jan was doing, can deepen students' understanding of arithmetic while simultaneously building mathematical habits of mind that support learning far into the students' mathematical future.

One subcategory is *treating algebra as generalized arithmetic*, with emphasis on the structural features of numbers and operations, including, for example, a focus on the special properties of zero and one, the generality of the result of subtracting a number from itself, the commutativity of addition and multiplication, the distributivity of multiplication over addition, and, more generally, the structural properties of the integers as an integral domain. (This aspect of algebraic reasoning is emphasized by Carpenter, Levi, Berman, & Pligge, chap. 4, this volume.)

Another subcategory is *exploring equality as expressing a relationship between quantities*. We coded eight occasions of exploring the algebraic character of "=" (all but one spontaneous). Jan spent time developing the notion of equality as a relationship between quantities by using a balance scale and problems such as 8 + 4 = ___ + 5 (Falkner, Levi, & Carpenter, 1999), a kind of problem explored in the seminars. In modeling problems with the scale, students worked on both sides of the equality to counter the pull of repeated computational exercises to interpret "=" as an action object signaling a computation result (Behr, Erlwanger, & Nichols, 1976; Kieran, 1983). As a result, they began to treat equations as objects expressing quantitative relationships. In one episode, Jan asked students to solve $(3 \times n) + 2 = 14$:

> Sam said that we could take the 2 away. He said that if we take the 2 from one side we have to take it from the other side. This was to make it balance. After we [took] the 2 away, he said to take the 12 tiles and put them in groups of 3. There were 4 groups, so the answer had to be 4. We tried replacing the "n" with 4 and it worked.

In Sam's strategy for solving the resulting equation $3 \times n = 12$, "dividing both sides by 3" was apparently not an operation available to him, and he could not perform equivalent operations on both quantities as he had previously done. Still, his alternative strategy of thinking of 12 as groups of 3 was not only quite sophisticated, but it also implicitly required him to see $3 \times n$ as equivalent to 12, simultaneously reorganizing one quantity in terms of another.

Algebraic Reasoning as the Construction and Use of Symbolic Forms as a Basis for Reasoning.   This cross-cutting category includes two subcategories: *writing formal expressions to express generalizations*, including those about patterns and functions; and using *unknowns and variables in equations*, especially in solving missing-number sentences. Writing and the use of symbolic forms typically occur in other categories of algebraic reasoning, reflecting the dual-sided nature of algebra and the importance of writing unexecuted arithmetic expressions.

In working with odd and evens, Jan led the students to write evens in the form "2 times n" (eventually abbreviated to $2n$). By noticing that any odd number could be thought of as following an even, they could write odds in the form $2n + 1$. Jan (and we) were surprised at students' willingness to write such expressions as abbreviations for conceptually meaningful number statements and relations. Recent work by Carraher, Schliemann, and Brizuela (Carraher, Schliemann, & Brizuela, in press; Schliemann, Carraher, & Brizuela, in press) suggests that this aspect of algebra might not be as difficult to adopt as previously thought. Missing-number sentences were also generated by students, sometimes spontaneously, in the course of other tasks. One interesting episode was "Zolan's" solution to a triangle puzzle. The triangle was subdivided into regions, some containing numbers and others empty; the regions had specified additive relationships (e.g., a fixed total), and the goal was to "complete the triangle" by finding all the missing numbers. Zolan spontaneously symbolized the problem by generating a set of equations $(4 + d = 7; 7 + a = 12; e + 4 = 5)$ to solve for the missing numbers.

Jan also combined categories of reasoning. For example, after asking students to use base-10 blocks to solve missing-number sentences, she extended the task, integrating multiple aspects of algebraic reasoning. After a discussion in which students shared strategies for solving the sentence "$14 = 6 + n$," Jan asked students to solve "$140 = 60 + n$," then "$1,400 = 600 + n$," thus superimposing pattern development on the missing-number activity. This second category of algebraic reasoning is at the heart of the use of unexecuted expressions and number sentences in the triangle problem described earlier, and plays an essential role in the ability to algebrafy arithmetic problems, as in the following illustration. Because students had learned the song "The 12 Days of Christmas," Jan developed an activity that explored the number of gifts received:

> How many gifts did your true love receive on each day? How many total gifts did she or he receive on the first 2 days? The first 3 days? The first 4 days? How many gifts did she or he receive on all 12 days? If the song was titled "The 25 Days of Christmas," how many total gifts would your true love have received?

This problem was mathematically related to others used in the professional development (e.g., counting embedded triangles), but an important generative property of Jan's practice was her capacity to recognize an algebraic opportunity and integrate it into her regular instruction.

Algebraic Reasoning as Functional Thinking Using Graphs, Tables, and Formulas.   This aspect of algebraic reasoning is perhaps the most frequently occurring one in the literature supporting algebraic reasoning in the early grades (e.g., NCTM, 2001). It includes the subcategory of *covariational and iterative reasoning*, where students create, through counting or measuring, organized sets of quantitative data and work to establish a generalization about relations among the data; and the subcategory of *organizing and representing data*. These aspects of algebraic reasoning, partly because they were emphasized in the professional development, were heavily represented in Jan's classroom, with 71 substantive instances occurring in 57 classes. Writing relationships in formal terms includes writing sequences of unexecuted expressions or number sentences. We include here an extended example based on the handshake problem discussed earlier because it illustrates the orchestration of several categories of algebraic reasoning being used in fruitful combination in Jan's class.

Mathematizing the act of shaking hands required students to define a "handshake" (some students wanted to count the number of "shakes"—the number of times hands went up and down). Students needed to understand the correspondence between a set of counted handshakes and its representing number, operate on consecutive numbers of handshakes to determine growth in quantity, and determine a total amount. Jan used this problem as a basis for algebraic reasoning. This problem was among the earliest used in the seminars, and instructors had not then emphasized the use of unexecuted expressions but focused on teachers' exploration of ways to render the problem sensible at their grade levels. The first time Jan visited the problem, she had groups of students build tables listing numbers of people and corresponding handshakes, up to 12 people, and paying attention to collection and display of data. Group size limited the physical enactment, forcing students to find an iterative method to answer the question.

Jan returned to the problem later in the year with a view to building sequences of number sentences. She initially posed the problem for groups of 6, 7, and 8. Through physical enactment and counting, aided by tallymarks, students again determined the number of handshakes in their own groups of 4 and 5. They discovered that if the groups' enactment shrank by one person each time someone shook hands with all the remaining people, they could count handshakes more easily and avoid double-counting. This led to sums of the form $5 + 4 + 3 + 2 + 1 = 15$ for a group of 6. They repeated this process for groups of 7 and 8 and put the collapsed data into a table. When Jan brought the class together, she asked the class to *write a number sentence* that would give the number of handshakes in a group of 12. She had planned this pedagogical maneuver, which pushed attention from the physical action and tables to the additive expressions and their forms. Students looked for patterns in the unexecuted sums that resulted:

Jan: If there were 12 people here and they were going to shake hands, what would you do?

Student: You could only shake 11 people's hands.

Jan: Why?

Student: Because he can't shake his own hand [the number sentence begins with 11 as opposed to 12].

Jan: So how would your number sentence change if there were 12 people?

Student: 11, 10, 9, 8, 7, 6, 5, 4, 3, 2, 1.

The structure of the physical was carried over to the structure of the inscriptions—the formal number sentences—which were then computed to answer the original question. Jan then asked about a group of 20, which led immediately to the sum of numbers from 19 to 1. Students had generalized not a number pattern of total handshakes, but the structure of the number sentences. They were reasoning from the *form* of the number sentences, treating the sentences as algebraic objects. One student noticed that if he paired the first and last terms, he got 20 as an answer. Jan quickly began drawing lines between pairs of opposite summands, $19 + 1$, $18 + 2$, and so on, helping the class use the structure of the sum to support a more efficient computation. She then asked, "How many 20s are there?" At this point, the students were again being asked to reason from the form of the inscriptions, treating them as objects to be operated on: Pair the summands, count the pairs, multiply that count by 20, then add on the unpaired number (10). (See Carpenter, Romberg, Smetzer-Anderson et al., 2004, for an annotated video episode.)

Jan did not push the students toward a closed-form quadratic-function description of the general case, although she did exploit the student-generated opportunity to add opposite pairs. Jan had moved from reasoning with functions to reasoning from the forms of expressions. This, coupled with her exploitation of the Twelve Days of Christmas, her superimposing multiple forms of algebraic reasoning in the missing-numbers example, and many others from her daily teaching and more recent work as a teacher-leader suggest a highly generative practice that we take as an ideal of algebrafied elementary mathematics teaching.

Impact on Student Achievement: A Comparison of Third Graders' Performance

Although qualitative analyses of our data showed a development in students' capacity to form, express, and justify generalizations (see, e.g., Blanton & Kaput, 2000), we turn here to a quantitative analysis of students' performance on items appearing on standardized tests. We administered 14 test items (10 multiple-choice items, 4 open-response items) selected from the fourth-grade MCAS (including non-algebra-strand items), to Jan's class of 14 third-grade students (a few were absent the first test day). We administered the same items to a second third-grade control class with comparable, but slightly higher SES from the same school. Analysis of students' responses, in both individual and partner settings, offers evidence that, *if given full opportunity to build algebraic reasoning*, they would develop abilities to form and express generalizations in their mathematical thinking in increasingly formal ways, abilities detectable on standard test items. Published data on these publicly released test items also enabled us to compare these very low SES students' achievement with that of students across the district and state.

Jan's students took the same exam on three days: first individually under standard MCAS conditions, then with a partner, and finally through whole-class discussion. Her students were also asked to write justifications for their answers on the first day. The second and third days' investigations were intended to help identify and isolate language issues. Reading for these students was a significant issue, given that the test was calibrated linguistically for fourth graders and more than half of Jan's students lived in homes in which English was not the primary language. Whole-class discussion clarified how students' algebra expertise interacted with specific items. Control students took the exam individually only, were given roughly the same amount of time, but were not asked to provide explanations. The second (experienced) instructor had not participated in our seminars and employed methods similar to those used in the district.

Item-Based Comparison of Classes.   A student's achievement on the spring 1999 fourth-grade MCAS was determined to be (a) "advanced" if the student scored at least 81% of possible points; (b) "proficient" if the student scored at least 67%; and (c) "needs improvement" if the student scored at least 41%. Below 41% was regarded as "failing." Ten of Jan's 14 third-grade students were at or above the "needs improvement" level; analysis of item scores shows that Jan's third-grade class performed approximately as well as fourth graders statewide and significantly better than district fourth graders. Item analysis shows that Jan's class outperformed the control group on 11 of the 14 items (see Fig. 5.3a), and significantly better (a = 0.05) on 4 items. The items on which the control group outperformed the experimental group (although not at a statistically significant level) were multiple choice (items 3, 9, 13). Because we have no written justifications from the control group, we cannot attribute the differences to any factor beyond chance. Jan's students' scored 22% on item 13, much lower than the 41% cut-off level and the control group's score of 50%. In the written justifications provided by 50% of Jan's students on this item, the most common error was adding numbers given in the problem or in the list of possible responses to determine an answer. Of the four who answered the problem correctly, one counted, one drew a model of divided oranges, one gave no written justification, and the other gave a response that was unintelligible. Of the 14 items, 7 (items 2, 3, 6, 7, 8, 11, 14) were deeply algebraic and required students to find patterns generated numerically and geometrically, understand whole-number properties, and identify unknown quantities in number sentences. The experimental class outperformed the control class on most of these (6 of 7).

Comparison With State and District Performance.   Jan's third-grade class performed approximately as well as the fourth graders statewide and significantly better than district fourth graders (whose performance had improved considerably over the prior year). Given the significant advantage of an additional year's instruction of the fourth-grade comparison groups, the significant development in verbal skills during the intervening year, and the low SES factors for the experimental class (particularly their linguistic backgrounds), we regard these as strong results. Nonetheless, we also know that certain features of students' algebraic reasoning competency

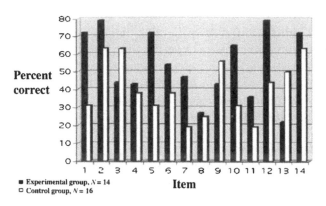

(a)

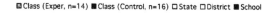

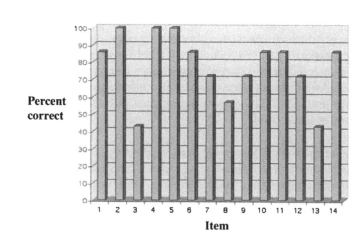

(b)

(c)

FIG. 5.3   Item analysis (selected MCAS items) (a) Comparison of scores: experimental and control groups. (b) Comparison of scores: experimental group, control group, and state, district, and school levels. (c) Scores by students working in pairs (experimental group).

were not tapped by the test items (e.g., their ability to write and reason with sequences of number sentences).

Item Analysis: Individual Students. Figure 5.3b provides an item analysis comparison of performances by the experimental and control groups at the state, district, and school levels on items selected from the 1999 MCAS (items 1, 2, 3, 4, 5, 6, 9, 10, 12, and 14). (Similar data for the other four items we used in our assessment were not available.) Again, we see that Jan's students performed comparably to students at the state, district, and school levels for seven items (items 1, 2, 5, 6, 10, 12, 14), exceeding fourth-grade results in items 10 and 12.

Item Analysis: Students Working in Pairs. As expected, students performed better with partners: "needs improvement" or better on 100% of the items, "proficient" or above on 79%, and "advanced" on 57% (see Fig. 5.3c). Five of the problems (items 1, 2, 4–7, 9–12, 14) on which students scored at the "proficient" or "advanced" levels were strongly algebraic.

Overall scores for pairs were quite strong. We believe that, beyond practice effects, this factored out the influence of verbal skills because the students were able jointly to interpret the problem wording, overcoming linguistic barriers. Out of 7 partner groups, all performed at least at the "needs improvement" level, with a maximum score of 90% and a minimum score of 47%; 81% (6 out of 7) performed at least at the "proficient" level; and 57% (4 out of 7) performed at the "advanced" level. For the most part, students took the assessment quite seriously and were deeply engaged in argumentation and justification with each other during the partner exam and the whole-class discussion—additional confirmation of the type of sociomathematical norms, critical for the development of students' algebraic thinking, that had evolved over the year.

POSTSCRIPT

The combination of student–teacher data we collected was intended to provide a picture of one attempt to come to terms with what we have described as the U.S. algebra problem, *a major national challenge,* and to provide early evidence that the teacher-development-centered solution strategy based on algebrafying elementary school mathematics is feasible—at a proof-of-concept level. Although we feel confident in asserting that elementary school teachers can algebrafy their practice given sufficient professional development and support, we have not yet addressed what it would take to implement this strategy on a massive scale. An approach based on the use of teacher-leaders in school-based professional development is currently under way. Further, the work by Franke, Carpenter, Levi, and Fennema (2001) and our own data on teacher growth suggest optimism regarding the long-term generativity of the algebrafying approach. In effect, we expanded a successful approach based on teacher knowledge of student thinking to an approach that applies that knowledge directly to teachers' instructional materials environment in ways that are inherently generative. As should be evident from the algebrafying activities described in this chapter, our professional development agenda is very ambitious: The needed pedagogical content knowledge reaches well beyond what teachers have experienced in their formal education and in most professional development. For most teachers, algebrafying will be a long-term process requiring long-term school-based support by on-site teacher-leaders.

We are confronting a problem of historical proportions that will become more serious as the mathematical needs of the population continue to grow in the 21st century. The universal mathematical objective from the 15th century to the latter decades of the 20th, was "shopkeeper arithmetic" (Swetz, 1987). The approach offered here recasts elementary mathematics in a profound way, not by ignoring its computational agenda, but by enlarging the agenda in ways that include the old in new forms that deliberately contextualize, deepen, and leverage the learning of basic skills and number sense by integrating them into the formulation of deeper mathematical understandings. We feel we have shown that, under appropriate support conditions, this approach is intelligible to elementary teachers in successfully actionable ways.

Finally, it is important to emphasize what we *did not* do. We did not attempt to develop elementary school versions of secondary school algebra as practiced in the United States or elsewhere. Our work did not involve teaching students or teachers simplified versions of the standard first algebra course or even preparing them for such algebra via specifically designed pre-algebra activities of the sort often used in the middle school. Rather, this work involved treating elementary school mathematics, especially but not exclusively arithmetic, in a new, more algebraic way. The approach was *transformative rather than additive*: It did not involve insertion of new curricular material into existing material, but treated core ideas and skills in ways that privileged certain activities and the expressions of generalization in increasingly formal and conventional ways, both of which we take to be at the heart of algebraic reasoning.

REFERENCES

Bednarz, N., Kieran, C., & Lee, L. (Eds.). (1996). *Approaches to algebra: Perspectives for research and teaching*. Boston: Kluwer.

Behr, M., Erlwanger, S., & Nichols, E. (1976). *How children view equality sentences* (Tech. Rep. No. 3). Tallahassee, FL: Florida State University. (ERIC Document Reproduction Service No. ED144802)

Blanton, M. L., & Kaput, J. J. (1994). *Characterizing generative and self-sustaining change in instructional practice that promotes algebraic reasoning.* Manuscript submitted for publication.

Blanton, M. L., & Kaput, J. J. (2000). Generalizing and progressively formalizing in a third-grade mathematics classroom: Conversations about even and odd numbers. In M. L. Fernández (Ed.), *Proceedings of the annual meeting of the North American Chapter of the International Group for the Psychology of Mathematics Education* (Vol 1, pp. 115–119). Columbus, OH: ERIC Clearinghouse.(ERIC Document Reproduction Service No. ED446945)

Blanton, M. L., & Kaput, J. J. (in press). Design principles for instructional contexts that support students' transition from arithmetic to algebraic reasoning: Elements of task and culture. In R. Nemirovsky, B. Warren, A. Rosebery, & J. Solomon (Eds.), *Learning environments: The encounter of everyday and disciplinary experiences.* Mahwah, NJ: Lawrence Erlbaum Associates.

Carpenter, T. P., Romberg, T. A., Smetzer-Anderson, S. et al. (2004). *Powerful practices in mathematics and science.* Madison, WI: National Center for Improving Student Learning and Achievement in Mathematics and Science; Naperville, IL: North Central Eisenhower Mathematics and Science Consortium at NCREL.

Carraher, D., Schliemann, A. D., & Brizuela, B. (in press). Treating operations as functions. In D. W. Carraher, R. Nemirovsky, & C. DiMattia (Eds.), Media and meaning [CD-ROM special issue]. *Monographs for the Journal of Research in Mathematics Education.*

Falkner, K. P., Levi, L., & Carpenter, T. P. (1999). Children's understanding of equality: A foundation for algebra. *Teaching Children Mathematics, 6*(4), 232–236.

Ferrini-Mundy, J., Lappan, G., & Phillips, E. (1997). Experiences with patterning. *Teaching Children Mathematics, 3*(6), 282–289.

Franke, M. L., Carpenter, T. P., Levi, L., & Fennema, E. (2001). Capturing teachers' generative change: A follow-up study of professional development in mathematics. *American Education Research Journal, 38*(3), 653–689.

Glenn, J., et al. (2000, September). *Before it's too late: A report to the nation from the National Commission on Mathematics and Science Teaching for the 21st century.* Jessup, MD: U.S. Department of Education.

Hamming, R. W. (1962). *Numerical methods for scientists and engineers.* New York: McGraw-Hill.

Kaput, J. (1998). Transforming algebra from an engine of inequity to an engine of mathematical power by "algebrafying" the K–12 curriculum. In National Council of Teachers of Mathematics and Mathematical Sciences Education Board (Eds.), *The nature and role of algebra in the K–14 curriculum: Proceedings of a national symposium* (pp. 25–26). Washington, DC: National Research Council, National Academy Press.

Kaput, J. (1999). Teaching and learning a new algebra. In E. Fennema & T. Romberg (Eds.), *Mathematics classrooms that promote understanding* (pp. 133–155). Mahwah, NJ: Lawrence Erlbaum Associates.

Kaput, J., & Blanton, M. (2001). Algebrafying the elementary mathematics experience. Part I: Transforming task structures. In H. Chick, K. Stacey, J. Vincent, & J. Vincent (Eds.), *The future of the teaching and learning of algebra: Proceedings of the 12th ICMI study conference* (pp. 344–351). Melbourne, Australia: University of Melbourne.

Kieran, C. (1983). Relationships between novices' views of algebraic letters and their use of symmetric and asymmetric equation-solving procedures. In J. C. Bergeron & N. Herscovics (Eds.), *Proceedings of the fifth annual meeting of the North American Chapter of the International Group for the Psychology of Mathematics Education* (Vol. 1, pp. 161–168). Montreal, Quebec, Canada: International Group for the Psychology of Mathematics Education.

Kilpatrick, J., Swafford, J., & Findell, B. (Eds.). (2001). *Adding it up: Helping children learn mathematics*. Washington, DC: National Academy Press.

Kline, M. (1972). *Mathematical thought from ancient to modern times*. New York: Oxford University Press.

Lacampagne, C., Blair, W., & Kaput, J. (Eds.). (1995). *The algebra initiative colloquium:* (Vols. 1–2). Washington, DC: U.S. Department of Education.

Lewis, C. (1981). Skill in algebra. In J. R. Anderson (Ed.), *Cognitive skills and their acquisition* (pp. 85–110). Hillsdale, NJ: Lawrence Erlbaum Associates.

Lins, R. (1992). Algebraic and nonalgebraic algebra. In W. Geeslin & K. Graham (Eds.), *Proceedings of the 16th annual meeting of the International Group for the Psychology of Mathematics Education, 2,* 56–63.

Moses, B. (1995). Algebra, the new civil right. In C. Lacampagne, W. Blair, & J. Kaput (Eds.), *The algebra colloquium: Vol. 2. Working group papers* (pp. 53–67). Washington, DC: U.S. Department of Education.

National Center for Education Statistics. (1996). *Pursuing excellence: A study of U.S. eighth-grade mathematics and science teaching, learning, curriculum, and achievement in international context*. Washington, DC: U.S. Government Printing Office.

National Center for Education Statistics. (1997). *Pursuing excellence: A study of U.S. fourth-grade mathematics and science achievement in international context*. Washington, DC: U.S. Government Printing Office.

National Center for Education Statistics. (1998). *Pursuing excellence: A study of U.S. twelfth-grade mathematics and science achievement in international context* (NCES Publication No. 98-049). Washington, DC: U.S. Government Printing Office.

National Council of Teachers of Mathematics. (2000). *Principal and standards for school mathematics*. Washington, DC: Author.

National Council of Teachers of Mathematics. (2001). *Navigating through algebra series*. Washington, DC: Author.

National Council of Teachers of Mathematics & Mathematical Sciences Education Board. (Eds.). (1998). *The nature and role of algebra in the K–14 curriculum*. Washington, DC: National Research Council, National Academy Press.

Romberg, T. A., & Kaput, J. (1999). Mathematics worth teaching, mathematics worth understanding. In E. Fennema & T. Romberg (Eds.), *Mathematics classrooms that promote understanding* (pp. 3–17). Mahwah, NJ: Lawrence Erlbaum Associates.

Ruopp, F., Cuoco, A., Rasala, S., & Kelemanik, M. (1997). Algebraic thinking: A professional development theme. *Teaching Children Mathematics, 3*(6), 326–329.

Schliemann, A. D., Carraher, D. W., & Brizuela, B. (in press). *Bringing out the algebraic character of arithmetic: From children's ideas to classroom practice*. Mahwah, NJ: Lawrence Erlbaum Associates.

Sleeman, D. H. (1984). An attempt to understand students' understanding of basic algebra. *Cognitive Science, 8,* 387–412.

Swetz, F. (1987). *Capitalism and arithmetic: The new math of the 15th century*. Chicago: Open Court.

Thompson, A., & Thompson, P. (1996). Talking about rates conceptually: A teacher's struggle. *Journal for Research in Mathematical Education, 27*(1), 2–24.

Thompson, P. W. (1993). Quantitative reasoning, complexity, and additive structures. *Educational Studies in Mathematics, 25*(3), 165–208.

Thompson, P., & Thompson, A. (1987). Computer presentations of structure in algebra. In J. Bergeron, N. Herscovics, & C. Kieran (Eds.), *Proceedings of the 11th annual meeting of the International Group for the Psychology of Mathematics Education, 1,* 248–254.

Wagner, S. (1981). Conservation of equation and function under transformation of variable. *Journal for Research in Mathematics Education, 12,* 107–118.

Wenger, R. (1987). Cognitive science and algebra learning. In A. H. Schoenfeld (Ed.), *Cognitive science and mathematics education* (pp. 217–251). Hillsdale, NJ: Lawrence Erlbaum Associates.

Yarema, C., Adams, R., & Cagle, R. (2000). A teacher's "try" angles. *Teaching Children Mathematics, 6*(5), 299–302.

Chapter 6

Statistical Data Analysis: A Tool for Learning

Kay McClain
Paul Cobb
Vanderbilt University

Koeno Gravemeijer
Freudenthal Institute and Vanderbilt University

In response to a Presidential directive and to the results of the Third International Mathematics and Science Study, the U.S. Department of Education and the National Science Foundation (1998) released the joint report "An Action Strategy for Improving Achievement in Mathematics and Science." The report documented that U.S. students, in comparison to students from higher achieving countries, begin to lose ground in their understanding of and interest in mathematics and science during the middle school grades. Along with algebra and geometry, data analysis was identified in the report as an area in which "far too many [students] have failed to develop a foundation [suitable for] the 21st century."

The repercussions of this finding are important: Data analysis provides a setting in which connections can be developed both between mathematics and other disciplines such as science and social studies, and within mathematics itself. Data analysis also plays an increasingly central role in both work-related activities and informed citizenship (G. W. Cobb & Moore, 1997; de Lange, van Reeuwijk, Burrill, & Romberg, 1993; National Council of Teachers of Mathematics [NCTM], 2000), as evidenced by the ever expanding use of computers within a society that places an increasing premium on statistical reasoning. This focus on statistical data analysis (which involves formulating and critiquing data-based arguments) has dramatic implications for the discourse of and debates over public policy and, thus, for democratic participation and power (G. W. Cobb, 1997). Cast in these terms, statistical literacy, which involves reasoning with data in relatively sophisticated ways, bears directly on both equity and participatory democracy.

The importance of statistics as a core area in mathematics and science is highlighted in recent reform documents. The NCTM (2000) *Principles and Standards for School Mathematics* included a detailed discussion of data analysis, statistics, and probability as central aspects of curricula for Grades 6–8. The authors argued that exploratory data analysis should be a precursor to formal statistical inference. These standards emphasized the importance of students developing the ability to make inferences and arguments based on patterns in the data. The importance of statistics to the teaching of science was similarly highlighted in the National Research Council (NRC, 1996) report *National Science Education Standards*, which outlined a vision of science for all students and focused on the teaching of science as inquiry: "Students begin with a question, design an investigation, gather evidence, formulate an answer to the original question, and communicate the investigative process and results" (p. 143). This characterization frames the process of data collection and data analysis as a central aspect of scientific inquiry. The interpretation of data and the communication of the conclusions drawn are activities clearly central to the learning of science, and the guiding image that emerges from both recommendations is one of students engaging in instructional activities in which they develop and critique data-based arguments (cf. Wilensky, 1997).

In this chapter, we provide an overview and analysis of our work in supporting middle school students' development of central statistical ideas. The research methodology that we employ while working in classrooms falls under the general heading of design research (cf. A. L. Brown, 1992; P. Cobb, 2000; Confrey & Lachance, 2000; McClain, 2002; Simon, 2000; Suter & Frechtling, 2000). Design research is characterized by cycles of design and analysis that involve both the "engineering of particular forms of learning and [the systematic study of] those forms of learning within the context defined by the means of supporting them" (P. Cobb, Confrey, diSessa, Lehrer, & Schauble, 2003, p. 9). The successive iterations of testing and revising conjectures about the means of supporting students' learning, inherent in design research, are highly interventionist and have as their goal the development of explanatory theories that inform future iterations of research, instructional design decisions, and classroom practice.

In the work discussed here, we focus on two design experiments and draw on multiple data sources, including videotape (from two cameras) and field notes of each class session and copies of students' work. Retrospective analysis involved continually testing and revising conjectures developed prior to and during the design experiment while working through the data chronologically. The resulting claims or assertions spanned the data set, yet remained empirically grounded in the details of specific episodes. We conducted the first design experiment in the first 12 weeks of the seventh-grade year and the second (with some of the same students) in the first 14 weeks of the eighth-grade year, accounting for a 9-month lapse. Throughout both design experiments, McClain assumed primary responsibility for teaching and was assisted on numerous occasions by Cobb. In the sections that follow, we outline the design and decision-making process of the research team; highlight the ways the iterative cycles of design research supported the learning not only of students and teachers, but also of researchers; and relate our decision making to the students' subsequent learning. We conclude by relating our classroom design efforts to our current collaborative efforts with teachers.

CLASSROOM DESIGN EXPERIMENT: GRADE 7

The seventh-grade design experiment involved 34 class sessions (29 students) over a 12-week period. Our overall goal for the design experiment was to support the students' development of increasingly sophisticated ways of analyzing univariate data as part of the process of developing effective data-based arguments. Our plan was to develop a single, coherent instructional sequence that tied together the separate, loosely related topics that typically characterize middle school statistics curricula. Focusing on the key notion of *distribution* enabled us to treat notions such as mean, mode, median, and frequency as well as others such as "skewness" and "spread-out-ness" as characteristics of distributions, and to view conventional graphs such as histograms and box-and-whiskers plots as ways of structuring distributions. The goal of supporting a shift from reasoning about data additively (in terms of absolute frequency) to reasoning multiplicatively (in terms of relative frequency) was inherent in the approach we formulated (cf. Harel & Confrey, 1994; Thompson, 1994; Thompson & Saldanha, 2000). In the subsequent eighth-grade design experiment, we planned to build on students' prior experience of analyzing univariate data sets so that they might come to view bivariate data sets as distributions within a two-dimensional space of values.

The Computer Tools

As we began to develop the instructional sequence, we were guided by the premise that the integration of computer tools was critical in supporting our mathematical goals. Students would need efficient ways to organize, structure, describe, and compare large data sets so that they could engage in genuine analysis of data, that is, looking for trends and patterns in the distribution of the data. This could best be facilitated by the use of computer tools, but we wanted to avoid creating tools that would offer either too much or too little support. This quandary is captured in the debate (often cast in terms of expressive and exploratory computer models; see Doerr, 1995) about the role of technologies in supporting students' understandings of data and data analysis. In the *expressive* approach, students are expected to recreate conventional graphs with only an occasional nudging from the teacher. In the *exploratory* approach, the students work with computer software that presents a range of conventional graphs, with the expectation that students will develop mature mathematical understandings of the graphs as they use them. The approach that we took offers a middle ground. We introduced tools and ways of structuring data designed to fit with students' current ways of understanding, while simultaneously building toward the relatively sophisticated ways of organizing data inherent in conventional graphs (Gravemeijer, Cobb, Bowers, & Whitenack, 2000).

The Grade 7 sequence involved two computer tools. The first was explicitly designed to enable students to investigate trends and patterns in data sets and to manipulate, order, partition, and otherwise organize small sets of data in a relatively routine way. Part of our design rationale was to support students' ability to analyze data, as opposed to their simply "doing something with numbers" (cf., McGatha, Cobb, & McClain, 2002). When data were entered, individual data values were shown as bars, the length of which signified the numerical value of the data point. A data set, therefore, was a set of parallel bars (of varying lengths) aligned with the axis. This tool made it possible for students to act on data in a relatively direct way, determining the number of data points within a selected interval and (using a value bar that could be dragged along the axis to partition data sets) estimating the mean or marking the median. Had we used commercially available data-analysis software, which typically offers only a selection of conventional graphs, students' manipulation of data in the ways we intended would not have been possible.

The second computer tool built on the first. The end points of the bars, which each signified a single data point in the first tool, were, in effect, collapsed down onto the axis so that the data set was inscribed as a collection of dots located on an axis (i.e., a line plot, as shown in Fig. 6.1a). This tool offered a range of ways to structure data, which, importantly, did not correspond to the conventional graphs typically available in commercial software packages. Instead, tool options resembled the ways in which students structure data when given the opportunity to develop their own approaches to conducting genuine analysis (cf. Hancock, 1995). Two of these options can be viewed as precursors to standard ways of inscribing data: (a) partitioning the data into four equal groups so that each group contains one fourth of the data (precursor to the box-and-whiskers plot), and (b) organizing data into groups of a fixed interval width so that each interval was the same size (precursor to the histogram). The three other options available to students, however, did not correspond to graphs typically taught in school: creating their own groups, partitioning the data into groups of a fixed size, and partitioning the data into two equal groups. The first and least sophisticated of these options involved simply dragging one or more bars to chosen locations on the axis in order to partition the data set into groups of points. The number of points in each group was shown on the screen and adjusted automatically as the bars were dragged along the axis (see Fig. 6.1b).

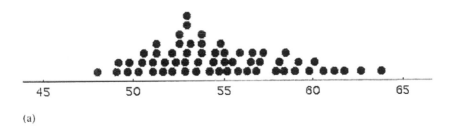

(a)

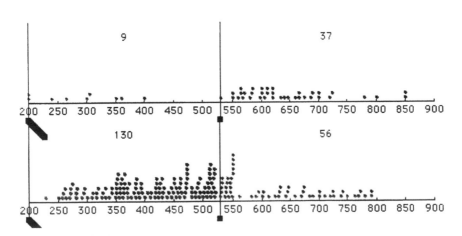

FIG. 6.1 Using the second computer tool. (a) Univariate data inscribed as line plots. (b) Create Your Own Groups option: Univariate data on T-cell counts of AIDS patients on two different treatment protocols, with cut point of 525.

Instructional Tasks

We reasoned that students should analyze data sets that they viewed as realistic for reasons that they considered legitimate. We anticipated that this could best be achieved by developing a sequence of tasks that involved either describing a data set or analyzing two or more data sets in order to make a decision, a judgment, or a specific, tailored recommendation (e.g., a recommendation to a chief medical officer about a better treatment for AIDS patients).

An important aspect of our instructional approach involved talking through the data creation process with the students. In situations where students did not themselves collect the data, we considered it important that they think about the decisions made in determining the data needed and the way they were collected. In class, the teacher engaged the students in lengthy discussions during which the students typically made conjectures and offered suggestions about the information needed to make a reasoned decision. In this process, students listed the criteria on which they would gather measurements (e.g., the response times of two competing ambulance companies). Against this background, they discussed the measures that they could take to collect adequate data for their subsequent decisions, in the process creating and then clarifying the design specifications for their "study." These discussions proved critical in grounding the students' data analysis in the context of a recommendation that had "real" consequences.

Our purpose in developing these instructional activities was to ensure that the students' reasoning had the spirit of genuine data analysis from the outset. The importance of attending to the authenticity of the instructional activities became apparent once we acknowledged that anticipation is at the heart of data analysis. Proficient analysts anticipate that certain ways of structuring and inscribing data might reveal trends, patterns, and anomalies that bear on the questions at hand. These anticipations reflect a deep understanding of central statistical ideas. For example, a student who decides that it might be productive to inscribe data sets as box plots anticipates that structuring the data sets in this way might lead to the identification of relevant patterns. Viewed in this way, the development of increasingly sophisticated ways of structuring and organizing data is inextricably bound up with the development of increasingly sophisticated ways of inscribing data (Biehler, 1993; P. Cobb, in press; de Lange et al., 1993; Lehrer & Romberg, 1996). The challenge, as we formulated it, was to transcend what Dewey (1981) termed the dichotomy between process and content by systematically supporting the emergence of key statistical ideas such as distribution while simultaneously ensuring that students' classroom activity was commensurate with the spirit of genuine data analysis. As Biehler and Steinbring (1991) noted, an exploratory or investigative orientation is central to data analysis and constitutes an instructional goal in its own right.

The decisions we made while preparing for the design experiment resulted in a conjectured learning trajectory (cf. Simon, 1995) of the students' learning, together with the conjectured means of supporting that learning. The research team developed a metalevel frame that offered an overall orientation and sense of directionality, a sequence of possible instructional tasks, and two accompanying computer-based tools for analysis. Although the intent of the sequence was outlined prior to the design experiment, decisions about specific tasks were made on a daily basis and guided by daily analyses of the students' classroom mathematical activity.

Characterizing Data.   Analysis of preassessment data had indicated that most of the students typically calculated the mean when comparing two sets of data. Our intent in these first tasks was to create a perturbation in students' current ways of thinking (e.g., a focus on measures of center) and initiate shifts toward ways of reasoning that focused on other features of data sets. The goal was that the students would come to reason about and characterize the data as they made comparisons across the data sets and found ways to address the question at hand.

One of the initial task situations involved comparing data on two brands of batteries: Always Ready and Tough Cell. Students framed the discussion around their experiences using batteries in electronic devices and reflected on how often they purchased batteries, noting that a better battery would be one that lasted longer. Students were then asked to analyze data from the testing of a sample of 20 batteries (10 each of two brands; see Fig. 6.2).

The students worked in pairs, using the first computer tool to organize and structure the data in ways that helped them make a decision. Afterward, they discussed the results of their analysis in a whole-class setting. Cynda began by explaining that she had used the range tool to identify the "top 10" of the 20 batteries tested, noting that 7 of the longest lasting were Always Ready batteries. During the discussion, Jason pointed out that if Cynda had chosen the top 14 batteries instead of the top 10, there would have been 7 of each battery. Brad then noted that he had compared the two brands differently: He had used the value bar to partition the data and had noted that all 10 of the Tough Cell batteries had lasted more than 80 hours, but that only 8 of the Always Ready batteries had. He noted, "I'd rather have a consistent battery that is going to give me above 80 hours instead of one where I just have to guess." When another student questioned his reasoning, noting that only two of the Always Ready batteries hadn't lasted that long, Brad argued that "the two or three will add up. It will add up to more bad batteries and all that."

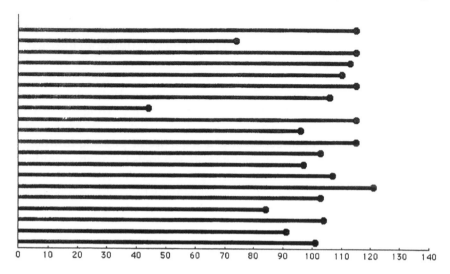

FIG. 6.2 Using the first computer tool: Data on the longevity of two brands of batteries (as presented to students).

Although most of the students understood what Cynda had done (i.e., she looked at the top half of the data and noted which brand had the most data points there), her choice of the "top 10" was arbitrary and, as Jason recognized, not valid for the investigation at hand. Brad's rationale, however, was grounded in the situation-specific imagery of the longevity of the batteries and seemed to make sense (i.e., to be valid) to most of the students. Brad wanted batteries that he could be assured would last a minimum of 80 hours. The discussion brought about an important shift in students' reasoning about data.

As students explored similar investigations, many used the value bar to partition the data, placing it at a particular value (typically not arbitrary) along the axis and then reasoning about the number of data points above or below that value. In one task, for example, as they investigated the ages at which U.S. citizens required increased health care, many students placed the value bar at 65 years of age, arguing that this enabled them to focus on senior citizens in comparison to the rest of the population. We should clarify that we did not anticipate that the students would use the value bar in this way. Our expectation when we designed the tool was that they might use the bar to estimate the mean of a data set. Instead, the students adapted this feature of the computer tool to their current ways of thinking about data.

Reasoning About Trends in Data.   We introduced the second computer tool once we judged that partitioning data sets to make data-based arguments had become normative and partitioning data sets was beyond justification in the analysis process. One of the initial investigations with the second computer tool involved analyzing data on driving speeds on a busy city road. The police department had decided to set up a speed trap to try to slow the traffic. Data were collected on the speeds of the first 60 drivers to pass the data collection point on a Friday afternoon before the speed trap was in place. The second set of data was collected at the same location, 4 weeks later. The intent of this task was to pose a situation that would create two data samples with equal numbers of data points and that could be examined by focusing on shifts in the distributions of the data.

Students began by discussing what information would be necessary to determine if the speed trap was effective. After much discussion on issues of safety related to speed and of the specifics of this particular problem, students were asked to compare the two sets of collected data to decide if the speed trap was effective (see Fig. 6.3).

After students had analyzed the data using the computer tool, we asked them to develop a written argument that could be submitted to the Chief of Police. In the subsequent whole-class discussion, students read their reports as part of their explanations.

The first argument was presented by Janice:

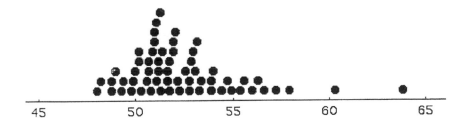

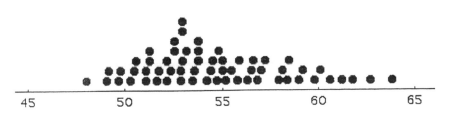

FIG. 6.3   Using the second computer tool: Data on speeds of drivers before (bottom) and after (top) a speed trap was put in place.

If you look at the graphs and look at them like "hills," then for the "before" group it is more spread out, and more are over 55. If you look at the "after" graph, then more people are bunched up closer to the speed limit, which means that the majority of the people slowed down.

After Janice had completed her explanation, the teacher used the projection system to display the data on the white board and explained:

Okay, Janice said if you look at this like hills ... now think about this as a hill [draws a hill over the first set] and think about this as being a hill [draws a hill over the second set], see what Janice was talking about? Before the speed trap, the hill was spread out, but after the speed trap, the hill got more bunched up, and [fewer] people were speeding.

From the ensuing discussion, we inferred that most of the students understood Janice's argument and saw the relevance of her idea of "hills," as indicated by Kent's comment:

They were slowing down. I want to compliment Janice on the way that she did that. I couldn't find out some way to compare, and I think that was a good way.

The key point for us is that the notion of a data set as a distribution of data points emerged as the students discussed Janice's "hills." She was concerned with distribution and focused on qualitative proportions (e.g., the majority). In this episode, many of the students began to reason about global trends in entire data sets as characterized by the "hills." This was the first occasion in which distribution became an explicit focus of discussion. Previously, students had focused on the number of data points in parts of data sets (as in the batteries task) instead of qualitative distinctions between data distributions.

Reasoning About and Comparing Relative Frequency in Different Data Distributions. A further shift in the students' reasoning was seen 6 days later. Although interpreting data sets multiplicatively had become normative at this point, many students based their arguments on qualitative proportions based on perceptual patterns in the data. We hoped to initiate a shift toward multiplicative ways of structuring data such as the use of four equal groups. In developing the task, the research team had reasoned that a situation in which the number of data points in the two data sets differed greatly would create a situation in which additive and multiplicative structures could be compared and contrasted.

In the task we developed, students were asked to analyze the T-cell counts of two groups of AIDS patients enrolled in different treatment protocols. A lengthy discussion revealed that the students were quite knowledgeable about AIDS and understood the importance of finding an effective treatment. Further, they clarified the relation between T-cell counts and a patient's overall health (i.e., increased T-cell counts are desirable). In the task, students were given data on the T-cell counts of 46 patients in a new, experimental treatment and the T-cell counts of 186 patients in a standard protocol (see Fig. 6.4). The students were asked not only to make a recommendation about which protocol was more effective, but also to develop inscriptions that could be used to support their arguments.

As the students worked at their computers in groups, we monitored their activity in order to select students whose arguments might provide opportunities for shifts in mathematical thinking to occur. In one of the first reports discussed, the students had partitioned the data at a T-cell count of 525 and found that most of the data on the standard protocol were below that cut point, whereas most of the data on the experimental protocol were above. During the discussion, the teacher clarified that these students had chosen the T-cell count of 525 because the "hill" of one data set was mostly below this value and the "hill" of the other was mostly above. Thus, the students had partitioned the data at a particular value in order to develop a quantitative description of a perceived qualitative difference between the two data sets. Toward the end of the discussion, Janice commented, "I think it would be helpful to know how many of the possible [patients] were in that range."

At a student's suggestion, the teacher drew a diagram showing the number of patients in each treatment with T-cell counts above and below 525 (see Fig. 6.4b):

Teacher: Hey, I've got a question for everybody. Couldn't you just argue rather convincingly that the old treatment was better because there were 56 people over 525—56 patients had T-cell counts greater than 525—and here there are only 37, so the old has just got to be better. I mean there are 19 more in there, so that's the better one surely.

Brad: But there is more in the old.

Jake: Thirty-seven is more than half of 9 and 37, but 56 is not more than half of 130 and 56.

Kent: I've got a suggestion. I don't know how to do it [inaudible]. Is there a way to make 130 and 156 compare to 9 and 37? I don't know how....

Kent's suggestion indicates that he wanted to find a way of comparing the information in the diagram, a process similar to stating design specifica-

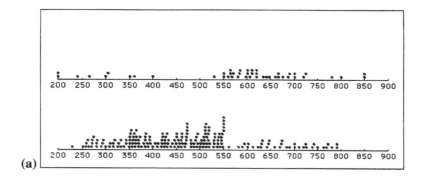

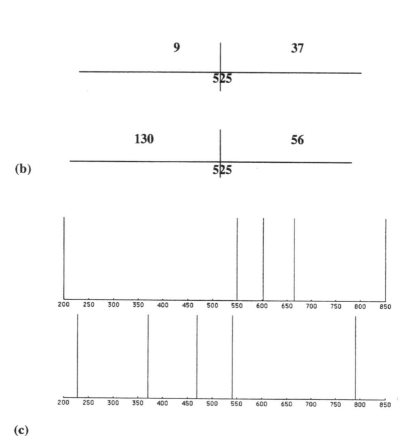

FIG. 6.4 Using the second computer tool: Inscription of univariate data on T-cell counts of AIDS patients on two different treatment protocols. (a) Line plots, options off. (b) Using Create Your Own Groups option, with cut point of 525 and individual data points hidden. (c) Using Four Equal Groups option, with the individual data points hidden.

138

tions for relative frequencies. Other students also questioned the teacher's additive argument. The teacher capitalized on this by introducing percentages as a solution. The students then calculated the percentages of patients in each treatment above and below a T-cell count of 525 and used the results to substantiate their initial arguments.

The teacher built from the use of percentages to introduce the next report in which the students used the computer tool to organize the data into four equal groups. The inscription they developed consisted of axes marked only with each of the resulting intervals, similar to a box-and-whiskers plot (see Fig. 6.4c). Brad noted that, with data partitioned this way, he could "tell where the differences is in the groups," and he thought "that the new treatment was better than the old treatment." When questioned about his reasoning, he explained, "Because the three lines for the equal groups [for the new treatment] were all above 525, compared to only one [line] on the old one." To Brad, this difference in intervals justified recommending the new treatment.

During the discussion, many students came to understand the need for multiplicative forms of structuring the data. We note that this episode also illustrates the contrast between constructing graphs by capitalizing on perceived trends in the data (e.g., "hills") and constructing graphs in order to read trends and patterns in the data. This additional shift most of the students were yet to make, but most did so within the first weeks of the second design experiment.

Findings

The metaphor of a hill (used to describe the shape of univariate data sets) emerged shortly after the students began using the second tool. By the end of the seventh-grade design experiment, most students understood that the data in a particular interval were a qualitative proportion (e.g., majority, most) of the entire data set rather than a mere additive part, indicating that most of the students had made the crucial transition from additive to multiplicative reasoning about data. As an example, in discussions of the AIDS data, students routinely spoke of the "majority" of the data, or "most of the people" and appeared to agree that "the majority" and "most" signified a qualitative proportion of a data set and, thus, a qualitative relative frequency.

The students who partitioned the data sets at T-cell counts of 525 did so because the "hill" in the experimental protocol data was above 525 whereas the "hill" in the standard treatment data was below 525. They were using the second computer tool to identify and describe perceptually based patterns in the data (e.g., hills or clusters). In contrast to this, a number of students compared the two treatment programs by using the computer tool to organize data sets independently of visual features. Consider again, for example, Fig. 6.4c, in which the AIDS data were partitioned using the Four Equal Groups option of the computer tool, with the individual data points hidden. In this case, the partitions did not isolate perceived clumps or hills in the data that could be read as qualitative proportions but served as a means of comparing and contrasting the ways in which the two sets of data were distributed.

A series of discussions conducted near the end of the teaching experiment indicated that a significant number of students were not yet able to infer distribution of data from a graph. This finding was substantiated by individual student interviews conducted immediately after the design experiment. All 29 students reasoned about univariate data in terms of qualitative proportions (i.e., qualitative relative frequency), but only 19 could use graphs of equal-interval widths (i.e., histograms) and of four equal groups (i.e., box plots), in which the data were hidden, to develop effective data-based arguments (P. Cobb, 1999).

By the end of the design experiment, students were able to compare data sets with unequal numbers of data values in terms of the proportion of data within various ranges of the data values (e.g., comparing the percentage of AIDS patients with T-cell counts above 525 instead of comparing the absolute frequencies). Konold, Pollatsek, Well, and Gagnon (1996) argued that a focus on the rate of occurrence of a set of data values within a range of values is at the heart of "a statistical perspective." Most of the students appeared well on the way toward developing this perspective.

We concluded from our analyses that the following had become normative: (a) using the metaphor of a "hill" to describe the shapes of data sets inscribed as line plots (a first step in discerning perceptually based patterns in the data), (b) comparing univariate data sets by structuring them in terms of perceptually based patterns (focusing on the shape or shift of distribution), and (c) reasoning multiplicatively about similarly structured data sets in terms of qualitative proportions (describing relative proportions of the data in terms of "majority" or "most." We also found the notion of distribution to be a viable idea around which to organize instruction at the middle school level, at least in the case of univariate data, and our conjectures about a learning route for students appeared to fit globally with the realized learning route of the students): (a) working out issues in creating data sets and characterizing data, (b) making the transition from additive to multiplicative reasoning as they come to view univariate data sets as distributions, and (c) in making this transition, becoming able to use histograms and box plots to compare and contrast *unequal* data sets.

CLASSROOM DESIGN EXPERIMENT: GRADE 8

The second design experiment began 9 months after the first and involved 41 lessons conducted over a 12-week period. The 11 eighth-grade students who participated in this experiment constituted a representative subgroup of the 29 who had participated in the seventh-grade design experiment. Our overall goal was to support the emergence of increasingly sophisticated ways of analyzing bivariate data and bivariate distributions as students worked to develop effective data-based arguments. Notions such as the direction and strength of the relationship between two sets of measures would then emerge as ways of describing distributions within a two-dimensional space of values (Wilensky, 1997).

Our reference to a two-dimensional space of values indicates the central role that we attribute to scatter plots as a way of inscribing bivariate data. The image of classroom discourse that we had in mind was that of students talking about scatter plots and referring to them as texts of situations from which the data were generated. We wanted students to anticipate that the aspects measured (as they generated their data) covaried in some way and that the nature of that covariation could be read from a scatter plot, and we wanted this process to become normative. We took into account that students frequently read graphs of this type diagonally rather than vertically, focusing on the distance of points from a line of best fit rather than on deviations in the y-direction (Clifford Konold, personal communication, July 23, 1998). For these students, the line of best fit, rather than the two sets of measure plotted orthogonally, constitutes the frame of reference. We conjectured that, in contrast, proficient data analysts in effect view the graph as organized into vertical slices, each of which can be seen as the (univariate) distribution of the measures of one quantity for an interval of values of the other.

Our claim was not, of course, that skilled readers of scatter plots consciously partition graphs into slices. Instead, we conjectured that the perceptual activity of skilled readers implicitly involves tracking the distribution of measures of the y-quantity as they scan the graph. This process of scanning down a scatter plot vertically rather than diagonally across is explicit in procedures for finding the line of best fit. As will become apparent, this view of a bivariate distribution as a distribution of univariate distributions strongly influenced the research team's decision about the design of third computer tool.

Given the 9-month gap, our first goal was to investigate whether students' statistical reasoning had regressed. The instructional activities used in the initial sessions were performance assessment tasks in which the students used the second computer tool to compare univariate data sets. Analysis of these sessions revealed that the students' reasoning about data had not regressed since the end of the first design experiment. There were also clear indications that a number of the students made significant progress within the first week of this design experiment. In particular, although only 8 of the 11 students participating had been able to use graphs of equal-interval widths (i.e., histograms) and of four equal groups (i.e., box plots) to develop effective data-based arguments by the end of the first design experiment, all 11 students routinely developed arguments of this type after the first few sessions of the second design experiment.

Our immediate goal when the students began to analyze bivariate data was to support the development of ways of inscribing the data. To this end, we planned to ask the students to develop a graph or a diagram of a bivariate data set that would enable them to make a decision or a judgment and to raise the issue of the extent to which their inscriptions enabled them to assess the variation in one of the measured quantities when the other changed. Only when the relevance of this criterion became normative, did we plan, if necessary, to introduce the scatter plot as a way of inscribing data that made it easier to address the question at hand.

As part of the design experiment, we also wanted to become normative the notion that bivariate data consist of the measures of two attributes of each of a number of cases. We conjectured that in the discussions of the students' analyses, students needed to develop ways of talking that referred explicitly to *cases*, rather than speaking solely in terms of *measures*. This, we reasoned, might support the view that each dot on a scatter plot signified a single case, with its two measures indicated by its location with respect to the axes.

The Third Computer Tool

Against this background, we introduced the third computer tool, in which bivariate data could be inscribed as a scatter plot. Using this tool, students could adjust the scales of the axes by changing the maximum and minimum values. In addition, if students (using the feature Dots) clicked on any data point, perpendiculars from the axes to the dot appeared. We anticipated that the use of this feature in whole-class discussions would aid the teacher in ensuring that discourse was about relationships between the two measures of each of a number of cases, rather than about a mere configuration of dots scattered between two axes. The individual data points could also be hidden, an option designed to support conversations in which trends and patterns in the distribution of data were to be inferred from graphs.

Beyond these features, the third computer tool offered four differing ways of organizing bivariate data. The Cross option (the two-dimensional correlate of the Create Your Own Groups option included in the second computer tool) divided the data display into four cells and showed the number of data points in each cell (see Fig. 6.5a). The students could drag the center of the cross to any location, thereby changing the size of the cells; the number of data points in each cell adjusted accordingly. The Grids option (the two-dimensional correlate of the Equal Interval Width option included in the second computer tool) allowed students to select from a pull-down menu of grids that ranged in size from 4-by-4 to 10-by-10 (see Fig. 6.5b). The selected grid was superimposed on the data display, with the number of data points in each cell shown. The Two Equal Groups option (the correlate of the Two Equal Groups option included in the second computer tool) partitioned the data display into 4 to 10 columns, or vertical slices, the widths of which divided the horizontal axis into equal intervals. Within each slice, the data points were partitioned into two equal groups, with the median and the extremes displayed. The Four Equal Groups option (the two-dimensional correlate of the Four Equal Groups option included in the second computer tool) was similar to the Two Equal Groups option, except that the data points within each slice were partitioned into Four Equal Groups (see Fig. 6.5c).

The crucial development that made it possible for the students to interpret scatter plots as bivariate distributions, rather than merely as configurations of data points, built directly on their prior learning in the first design experiment. Initially, the students viewed each of the vertical stacks merely as a collection of data points that occupied an interval of values defined by the maximum and minimum values. Most of the students, however, could interpret a data stack in terms of shape when the Grids or Four Equal Groups options were used to organize the data. In the case of stacked data organized using the Grids option, for example, all students could trace the shape of individual data stacks as hills by reading from the graph where the data bunched up in each stack. In doing so, they interpreted each data stack as a univariate distribution rather than as a collection of data points that occupied an interval. A plot of stacked data, therefore, became a distribution of univariate distributions. As a consequence, when the students traced a line through what they referred to as "the hills," they were indicating a conjectured relationship of covariation about which the data were distributed.

Interpreting Bivariate Distributions

As we began the design experiment, analysis of the previous design experiment guided our initial decision-making process. We intended to build from the students' conceptions to support their understandings of bivariate distributions. Initial tasks in this sequence involved students defining as-

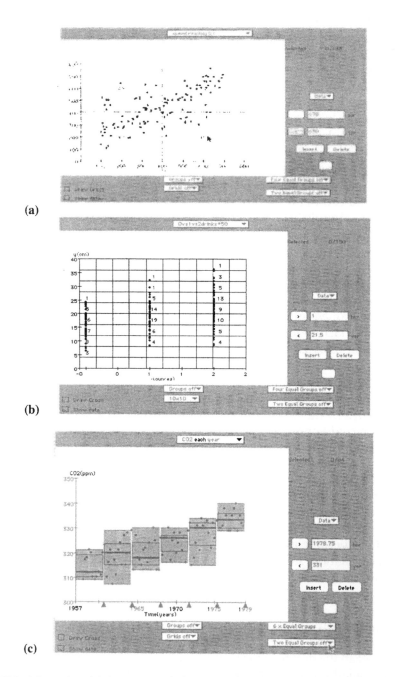

FIG. 6.5   Using third computer tool: Inscription of bivariate data. (a) Cross option. (b) Grids option. (c) Four Equal Groups option.

pects of patterns or trends in one data set (e.g., year and atmospheric CO_2 levels). In these instances, the students were able to discern trends and make predictions based on their analyses. We recognized, however, that they were able to do this without attending to the specificity of the trend. In one activity, for example, students noted that as the years went by, the CO_2 levels increased. They even made predictions about future CO_2 levels (5 years later) based on the trend. But when the teacher tried to focus the students' attention on particular features of the trend, the teacher was often unable to generate a need for such specificity. As a result, the students typically did not see merit in the question.

In response to this, one of the last tasks we posed to the students was intended to challenge students to move past simply characterizing trends to more clearly articulating features of the trends. The two data sets (120M, 120F) included data on the number of years of education and salary (see Fig. 6.6). The question was to determine if gender inequity was apparent in salary. Because both sets showed a positive correlation (i.e., the more education, the more money earned), an adequate analysis, then, required that the students tease out the salient aspects of those trends to highlight differences both within and across the data sets (e.g., greater skewness of the distribution in the women's salaries). Although in the process students would note that an ascending trend associated with increasing education was a factor associated with determining pay, simply stating that such a trend occurred would not be sufficient to answer the question.

The teacher began by asking students if they thought men made more money than did women. The class engaged in an extended discussion about issues of discrimination and speculated about circumstances that could contribute to inequity in salaries. The teacher then asked the students what information they thought they needed to investigate the question. In the ensuing discussion, students recognized that they needed data not only on the salaries, but also on factors that contributed to determining pay. Several students raised concerns related to sampling techniques. In response, the teacher clarified that the data they were using were taken from a random sample of IRS tax forms. The students not only questioned the use of the forms as a source of information, but also pointed to potential problems if samples were not also matched on experience, profession, age, and so forth (i.e., stratified random samples). As the conversation continued, several also questioned the relatively small sample size and argued that each data value represented "millions of people." As a result, they questioned whether their analysis legitimately addressed the question. After resolving these and a number of other issues related to the data creation process, the teacher asked the students to use the data available to see if, *based on these data*, they thought that men made more money. We had conjectured that the students would structure the data using Grids or Equal Groups options

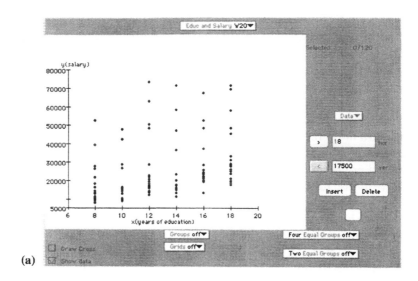

(a)

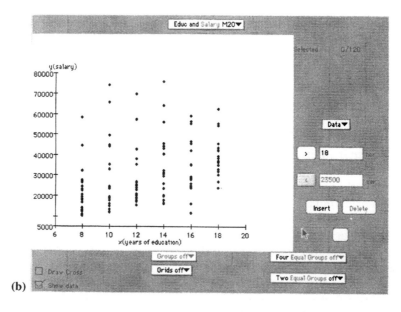

(b)

FIG. 6.6   Using the third computer tool: Salary and education (bivariate) data inscribed as stacked data (options off). (a) Women. (b) Men.

and reason about the changes in the distributions across the stacks. This would require their viewing the stacks as univariate distributions, reading trends and patterns across these distributions, and analyzing the data in two ways: comparing changes in salary with years of education *within* the data sets and comparing salary with the same years of education *across* the two data sets.

As the students worked at the computers, the teacher and members of the research team circulated among the groups. We noticed that many students initially focused on the extreme data values in each stack as well as on the medians. In earlier analyses, the students had often based their analysis on the extreme data values because they wanted to be able to describe the range within which all the data points fell. They used features on the computer tool to "capture" the data set inside the extreme values in order to reason about changes. We viewed this way of reasoning as highly problematic in that extreme values are typically unstable across samples and do not provide a basis for prediction and inference. In particular, we hoped to initiate a shift in the students' reasoning from a focus on selecting a value, to viewing the stacks as distributions. To support this, we worked on issues of sampling, asking students to note which values they might be able to predict across samples. Our goal was to build an understanding of the stability of the median across samples and a focus on the portion of the data clustered around this value. We anticipated that this, in turn, would contribute to the students' coming to view the median as a characteristic of a distribution. Nonetheless, several of the students still relied on the extreme values as they began their analyses of the salary data.

As the students worked, however, several began to speak of the extreme values as the exceptions (i.e., the people who made atypically high salaries) and focused more on the dense portions of the stacks (i.e., the data clustered around the medians). The median indicated where the cluster or "hill" of the data was located, supporting students' shift to a focus on the shape of the data. Students routinely spoke of the "majority" or "most of the data" as being located within the hill. These terms appeared to signify a qualitative proportion of a data set and, thus, a qualitative relative frequency. As they analyzed the data, most traced the shape of the hill in the stacks. We found this significant because the students and the teacher had previously drawn hills on data sets structured in Four Equal Groups to indicate the patterns of relative density in data stacks. Here, the students used this idea to frame their argument. As a result, the teacher decided to highlight these solutions in the subsequent whole-class discussion, making the specific characteristics of the distributions of the stacks the focus of analysis. This allowed students the opportunity to define the characteristics of the trends necessary to make a valid judgment, paying attention to trends not only within each data set, but also across the two data sets.

Importantly, all the students reached the same conclusion as a result of their analysis: They all agreed that, based on their data, men made more money. The norms that had become instantiated in the classroom, however, obliged the students to critique analyses in whole-class discussion in order to assess their adequacy in justifying the conclusion. As a consequence, whole-class discussions focused on the specific ways of structuring and organizing the data that would *best* support the conclusion, rather than on determining the course of action to recommend. It was, in fact, not unusual for most of the students to agree on the outcome of an analysis and still engage in discussions for an entire class period.

The teacher began the discussion by asking Brad and Mike to share their analysis of the distribution of the data in the stacks they referred to as "humps." The teacher chose this solution to discuss first because the students had focused on a perceived perceptual pattern in the univariate distributions or the stacks. Brad and Mike began by using the computer system to project the data sets onto the white board. They explained that they had initially looked at the extreme values in each stack and across the two data sets, but had not found this particularly useful. As Brad noted:

> We were, like, using the extremes at first, like, the men's highest and from the women the highest, but those don't really matter as much as, like, the medians and the humps because that's where, like, most people are, not where, like, just the exceptions of people who, like, maybe broke it big or whatever … are making that much money, but we, like, decided to go where most of the people are.

They then reasoned about the stacks of data by focusing on the "humps," clusters or clumps of data. The students listening both understood and agreed with Brad and Mike's analysis. The teacher, however, commented that "Nobody asked them a question, and I've never heard anybody use [hump] before, but you all seem to know what they are talking about."

Brad explained that there was usually a "hump" around the median, where the majority of the data were located. He then referred to the earlier class discussion in which the students and the teacher had drawn distribution curves on several stacks to clarify the shape of the data. As the discussion continued, Brad and Mike used the Two Equal Groups option to locate the medians (see Fig. 6.7). They pointed out to the class that the humps were located around the median, noting that the men's salaries (using the medians) were consistently higher than the women's salaries across the years of education: "See, you can see where the medians and humps are." Brad and Mike had identified a trend both within and across the data sets.

Sue, the next student asked to present an analysis, had reasoned about the data in a similar manner, but had structured the data sets using Four

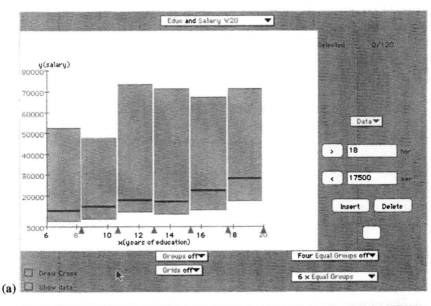

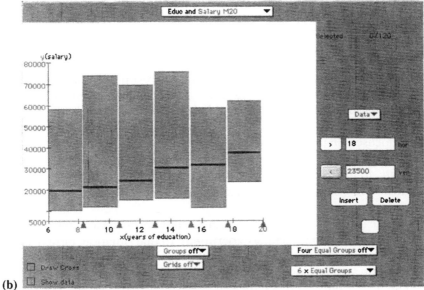

FIG. 6.7   Using the third computer tool: Salary and education (bivariate) data inscribed using Two Equal Groups option. (a) Women. (b) Men.

Equal Groups (see Fig. 6.8) and reasoned about the middle 50% of the two stacks because "the hump, or hill, or whatever, is in the middle 50%." After clarifying that the students understood Sue's way of reasoning, the teacher asked whether they thought this way of organizing the data was useful. Students stated it was important to know "what most people are making" and that focusing on the "middle" was a good way to do this.

At this point, the teacher built on Sue's presentation by asking the students to describe the trend in each of the two data sets. They agreed that salary increased with years of education for both the men and the women, but several noted that the *rate* of increase was greater for the men than for the women. To substantiate this observation, they used the computer tool to read the shape of the stacks and then contrasted the overall patterns in the two data sets (e.g., the skewness of the two distributions). We note that in doing so, they were not collapsing the data sets to lines, but were instead comparing trends in two bivariate distributions.

Findings

At first glance, scatter plots might seem relatively transparent, even to novices, as data involving two variables inscribed in a two-dimensional space of values. A cursory comparison with inscriptions of univariate data, however, indicates the fallacy of this argument. The line plot, for example, involves a second dimension, which (to a proficient user) indicates relative frequency. Scatter plots, however, do not provide such direct perceptual support for a third dimension corresponding to relative frequency. Instead, proficient analysts apparently read this third dimension from the relative density of the data points. Students incorporated this missing third dimension when they focused on the shape of data stacks and spoke of "hills" in two-dimensional data displays. This interpretation of stacked data in terms of relative density, in turn, enabled the students to use Grids and Four Equal Groups to structure what might be termed *unstacked data* into a distribution of univariate distributions.

These two ways of structuring data can be contrasted with the one we witnessed earlier in the teaching experiment. In that situation, most of the students focused on the extreme values of data stacks and seemed to view each stack as an amorphous collection of data points that occupied the space between these values. At that point, the stacks had little structure and were not univariate data distributions. As we have noted, the use of the hill metaphor enabled the students to view the stacks as a sequence of univariate distributions, to determine (in turn) trends and patterns across stacks, and, thus, to view the entire data set as a bivariate distribution. We note, however, that the lines that students sketched on scatter plots when they first used the

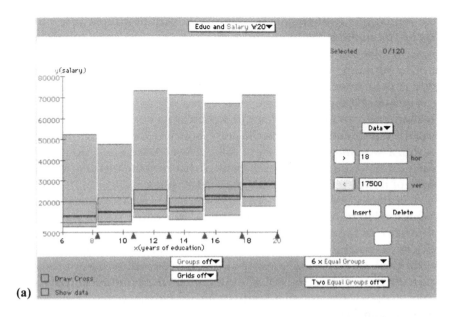

(a)

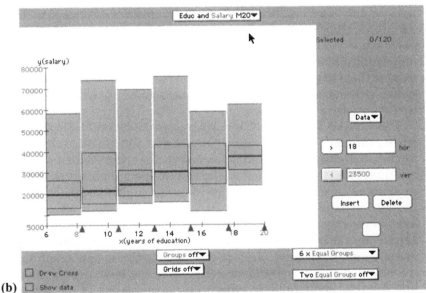

(b)

FIG. 6.8   Using the third computer tool: Salary and education (bivariate) data inscribed using Four Equal Groups option. (a) Women. (b) Men.

third tool appeared to be proposals about ways to summarize a dispersed configuration of data points by collapsing them down to a single line. At that point in the experiment, most of the students were not doing statistics in any real sense. Rather than developing ways to manage uncertainty, they were substituting certainty for variation. In contrast, our analyses indicate that in the final sessions of the second design experiment, most were interpreting lines of best fit (traced on scatter plots) as a conjectured relationship between the two variables about which the data were distributed. In other words, the students had come to view bivariate data sets as distributions within a two-dimensional space of values. As a consequence, notions such as direction and strength of the relationship between the variables emerged as ways of describing how the data were distributed.

We also note that, throughout the discussion of the IRS data, the students continued to raise questions about the validity of their analyses by making reference to the process that generated the data. In particular, when the teacher asked the students if, based on these data, they could make any predictions about men's and women's salaries for the entire country, they again raised concerns about the size of the sample and the possibility that the data might not be representative. Students also noted that years of experience in a job had not been accounted for in any way in the data set. We viewed these questions and issues as strong indicators that most of the students were developing a relatively deep understanding of data creation and its implications for the conclusions that could legitimately be drawn from the subsequent analysis. Students were able to step back from their analyses and critique them in light of the appropriateness of the data sampling and collection techniques as well as to question the validity of predictions and inferences made from such analyses.

The analyses also reveal that our conjectures about the effectiveness of talking through the data creation process were well founded. Initially, the teacher had to take a highly proactive role in guiding discussions of the data creation process, and the handover of responsibility from the teacher to the students was gradual. Eventually, however, the students began to anticipate possible limitations in the data creation process that might affect the legitimacy of the conclusions they could draw. During these discussions, for example, the students became increasingly concerned with controlling extraneous variables and raised questions related to issues of sampling and the representativeness of the data (e.g., as in the salary tasks), suggesting that they had come to understand that the legitimacy of the conclusions drawn from data depends crucially on the data generation process (see G. W. Cobb, 1997). This treatment of data as measures of an aspect of a situation rather than as mere numbers without context was maintained throughout both design experiments (Tzou & Cobb, 2000).

CONCLUSION

In this chapter, we focused not only on our findings concerning students' learning of statistical data analysis, but also on both the process of the students' learning and the specific means by which it was supported. These findings are pragmatically significant in that they provide the rationale for the instructional sequences and computer tools that we developed.

Our current work with middle school teachers is focused on supporting the emergence, development, and sustaining of professional teaching communities. Fundamental to this effort is a focus on the development of the teachers' content knowledge. We find that the learning trajectory presented here can serve not only as a conjectured route through mathematical terrain but also as a basis for guiding and supporting the development of the teachers (see McClain, 2003). During our collaboration with teachers, this conjecture is continually tested and refined. This approach includes taking not only the students' diverse ways of reasoning as a basis for the learning route of the teachers, but also the accompanying *means of support* as tools for supporting the evolution of the teachers' learning. These tools include the choice of tasks, the use of computer-based tools for analysis, the use of the teachers' inscriptions and solutions, and the norms for argumentation. Central to these efforts is a focus not only on the diversity of students' strategies and solutions, but also on their relative sophistication with respect to the mathematical agenda. Only in understanding *their students'* ways of reasoning can teachers support their students' development and achievement.

We also seek to outline an analytic approach that enables us to view teaching as a distributed activity and to situate teachers' instructional practices within the institutional settings of the schools and school districts in which they work. We know both from first-hand experience and from a number of more formal investigations that teachers' instructional practices are profoundly influenced by the institutional constraints that they attempt to satisfy, the formal and informal sources of assistance on which they draw, and the materials and resources that they use in their classroom practice (Ball, 1996; Brown, Stein, & Forman, 1996; Feiman-Nemser & Remillard, 1996; Nelson, 1999; Senger, 1999; Stein & Brown, 1997). In this work, we focus on (a) the role of activities in which members of different communities are jointly engaged, (b) the role of people who are (at least peripheral) members of two or more communities, and (c) the role of "objects" (i.e., tools, artifacts) incorporated into the practices of two or more communities (for more detail, see P. Cobb, McClain, Lamberg, & Dean, 2003). We pay particular attention to these interconnections because the use of tools and artifacts is a relatively inconspicuous, recurrent, and taken-for-granted aspect of school life, which is also underdeveloped in the research literature both on teacher professional development (Marx, Blumenfeld, Krajcik, & Soloway, 1998; Putnam & Borko, 2000) and on policy and educational leadership (Spillane, Halverson, & Diamond, 1999).

Our work attempts to account for the interpretations that teachers and other persons make and the understandings that they develop, to situate people's actions within the school or district as a lived organization, and to support the formulation of strategies for institutional change that involve the creation of new tools and practices, the orchestration of encounters between communities, and the development of a stronger role for people who are members of more than one community. This relatively broad orientation to the process of renegotiating the institutional settings in which teachers work is significant, given Wenger's (1998) observation that mutual engagement and reification offer two complementary ways of attempting to shape the future and suggest that one is rarely effective without the other.

We note, however, that our overarching goal in our analyses is motivated by pragmatic concerns. Ongoing analyses of the institutional setting in which the collaborating middle school teachers developed their instructional practices feeds back to inform our work with them. This analytic approach enables us to be more effective in collaborating with teachers in that it involves the development of testable conjectures about the constraints and affordances of the institutional settings in which teachers develop and revise their instructional practices. We contend that teachers' developing understanding of the relations between the institutional setting in which they work and their instructional practices is a crucial aspect of their learning, but note that it can also have a great effect on student achievement in areas beyond statistical reasoning.

AUTHOR NOTE

The research team for the seventh- and eighth-grade design experiments included the authors, Maggie McGatha, Lynn Hodge, Jose Cortina, Carla Richards, Carrie Tzou, and Nora Shuart. Cliff Konold and Erna Yackel were consultants on the project. The research team for the current work with teachers includes Cobb, McClain, Teruni Lamberg, Chrystal Dean, Jose Cortina, Qing Zhao, Jana Visnovska, Lori Tyler, and RaeYoung Kim.

REFERENCES

Ball, D. (1996, March). Teacher learning and the mathematics reforms: What we think we know and what we need to learn. *Phi Delta Kappan*, 500–508.

Biehler, R. (1993). Software tools and mathematics education: The case of statistics. In C. Keitel & K. Ruthven (Eds.), *Learning from computers: Mathematics education and technology* (pp. 68–100). Berlin: Springer-Verlag.

Biehler, R., & Steinbring, H. (1991). Entdeckende Statistik, Strenget-und-Blatter, Boxplots: Konzepte, Begrundungen, und Enfahrungen eines Unterrichtsver-suches [Explorations in statistics, stem-and-leaf, box plots: Concepts, justifications, and experience in a teaching experiment]. *Der Mathematikunterricht, 37*(6), 5–32.

Brown, A. L. (1992). Design experiments: Theoretical and methodological challenges in creating complex interventions in classroom settings. *Journal of the Learning Sciences, 2,* 141–178.

Brown, C., Stein, M., & Forman, E. (1996). Assisting teachers and students to reform the mathematics classroom. *Educational Studies in Mathematics, 31,* 63–93.

Cobb, G. W. (1997). More literacy is not enough. In L. A. Steen (Ed.), *Why numbers count: Quantitative literacy for tomorrow's America* (pp. 75–90). New York: College Entrance Examination Board.

Cobb, G. W., & Moore, D. S. (1997). Mathematics, statistics, and teaching. *American Mathematical Monthly, 104,* 801–823.

Cobb, P. (1999). Individual and collective mathematical development: The case of statistical data analysis. *Mathematical Thinking and Learning, 1,* 5–44.

Cobb, P. (2000). Conducting teaching experiments in collaboration with teachers. In A. E. Kelly & R. A. Lesh (Eds.), *Handbook of research design in mathematics and science education* (pp. 307–334). Mahwah, NJ: Lawrence Erlbaum Associates.

Cobb, P. (in press). Modeling, symbolizing, and tool use in statistical data analysis. In K. Gravemeijer, R. Lehrer, B. van Oers, & L. Verschaffel (Eds.), *Symbolizing and modeling in mathematics education.* Dordrecht, Netherlands: Kluwer.

Cobb, P., Confrey, J., diSessa, A., Lehrer, R., & Schauble, L. (2003). Design experiments in educational research. *Educational Researcher, 32*(1), 9–13.

Cobb, P., McClain, K., & Gravemeijer, K. (in press). Learning about statistical covariation. *Cognition and Instruction.*

Cobb, P., McClain, K., Lamberg, T., & Dean, C. (2003). Situating teaching in the institutional setting of the school and school district. *Educational Researcher, 32*(6), 13–24.

Cobb, P., & Whitenack, J. (1996). A method for conducting longitudinal analysis of classroom videorecordings and transcripts. *Educational Studies in Mathematics, 30,* 213–228.

Confrey, J., & Lachance, A. (2000). Transformative reading experiments through conjecture-driven research design. In A. E. Kelly & A. Lesh (Eds.), *Handbook of research design in mathematics and science education* (pp. 231–266). Mahwah, NJ: Lawrence Erlbaum Associates.

de Lange, J., van Reeuwijk, M., Burrill, G., & Romberg, T. (1993). *Learning and testing mathematics in context. The case: Data visualization.* Madison: National Center for Research in Mathematical Sciences Education, Wisconsin Center for Education Research.

Dewey, J. (1981). Experience and nature. In J. A. Boydston (Ed.), *John Dewey: The later works, 1925–1953* (Vol. 1). Carbondale: Southern Illinois University Press.

Doerr, H. M. (1995, April). *An integrated approach to mathematical modeling: A classroom study.* Paper presented at the annual meeting of the American Educational Research Association, San Francisco, CA.

Feiman-Nemser, S., & Remillard, J. (1996). Perspectives on learning to teach. In F. Murray (Eds.), *The teacher educator's handbook* (pp. 63–91). San Francisco: Jossey-Bass.

Franke, M. L., Carpenter, T. P., Levi, L., & Fennema, E. (2001). Capturing teachers' generative change. *American Educational Research Journal, 38,* 653–689.

Glaser, B., & Strauss, A. (1967). *The discovery of grounded theory.* Chicago: Aldine.

Gravemeijer, K., Cobb, P., Bowers, J., & Whitenack, J. (2000). Symbolizing, modeling, and instructional design. In P. Cobb, E. Yackel, & K. McClain (Eds.), *Symbolizing and communicating in mathematics classrooms.* Mahwah, NJ: Lawrence Erlbaum Associates.

Hancock, C. (1995). The medium and the curriculum: Reflections on transparent tools and tacit mathematics. In A. A. diSessa, C. Hoyles, R. Noss, & L. Edwards (Eds.), *Computers and exploratory learning* (pp. 49–69). Heidelberg: Springer-Verlag.

Harel, G., & Confrey, J. (1994) The development of multiplicative reasoning in the learning of mathematics. New York: State University of New York Press.

Konold, C., Pollatsek, A., Well, A., & Gagnon, A. (1996, July). *Students' analyzing data: Research of critical barriers.* Paper presented at the roundtable conference of the International Association for Statistics Education, Granada, Spain.

Lehrer, R., & Romberg, T. (1996). Exploring children's data modeling. *Cognition and Instruction, 14,* 69–108.

Marx, R. W., Blumenfeld, P. C., Krajcik, J. S., & Soloway, E. (1998). New technologies for teacher professional development. *Teaching and Teacher Education, 14,* 33–52.

McClain, K. (2002). A methodology of classroom teaching experiments. In S. Goodchild & L. English (Eds.), *Researching mathematics classrooms: A critical examination of methodology* (pp. 91–118). Westport, CT: Praeger.

McClain, K. (2003, April). *Tools for supporting teacher change: A case from statistics.* Paper presented at the annual meeting of the American Education Research Association, Chicago, IL.

McGatha, M., Cobb, P., & McClain, K. (2002). An analysis of students' initial statistical understandings: Developing a conjectured learning trajectory. *Journal of Mathematical Behavior, 16*(3), 339–355.

National Council of Teachers of Mathematics. (2000). *Principles and standards for school mathematics.* Reston, VA: Author.

National Research Council. (1996). *National science education standards.* Washington, DC: National Academy Press.

Nelson, B. (1999). *Building new knowledge by thinking: How administrators can learn what they need to know about mathematics education reform* (Paper 11). Center for Development of Teaching paper series. Newton, MA: Educational Development Center.

Putnam, R. T., & Borko, H. (2000). What do new views of knowledge and thinking have to say about research on teacher learning? *Educational Researcher, 29*(1), 4–15.

Senger, E. (1999). Reflective reform in mathematics: The recursive nature of teacher change. *Educational Studies in Mathematics, 37,* 199–221.

Simon, M. A. (1995). Reconstructing mathematics pedagogy from a constructivist perspective. *Journal for Research in Mathematics Education, 26,* 114–145.

Simon, M. A. (2000). Research on the development of mathematics teachers: The teacher development experiment. In A. E. Kelly & R. A. Lesh (Eds.), *Handbook of research design in mathematics and science education* (pp. 335–359). Mahwah, NJ: Lawrence Erlbaum Associates.

Spillane, J. P., Halverson, R., & Diamond, J. B. (1999, April). *Distributed leadership: Towards a theory of school leadership practice*. Paper presented at the annual meeting of the American Educational Research Association, Montreal, Quebec, Canada.

Stein, M. K. & Brown, C. A. (1997). Teacher learning in a social context. In E. Fennema & B. Scott Nelson (Eds.), *Mathematics teachers in transition* (pp. 155–192). Mahwah, NJ: Lawrence Erlbaum Associates.

Suter, L. E., & Frechtling, J. (2000). *Guiding principles for mathematics and science education research methods: Report of a workshop*. Washington, DC: National Science Foundation.

Thompson, P. W. (1994). The development of the concept of speed and its relationship to concepts of speed. In G. Harel & J. Confrey (Eds.), *The development of multiplicative reasoning in the learning of mathematics* (pp. 179–234). New York: State University of New York Press.

Thompson, P., & Saldanha, L. (2000). *To understand post counting numbers and operations*. White paper prepared for the National Council of Teachers of Mathematics.

Tzou, C., & Cobb, P. (2000, April). *Supporting students' learning about data creation*. Paper presented at the annual meeting of the American Educational Research Association, New Orleans, LA.

U.S. Department of Education and National Science Foundation. (1998). *An action strategy for improving achievement in mathematics and science*. Retrieved August 3, 2003, from http://www.ed.gov/pubs/12TIMSS/index.html

Wenger, E. (1998). *Communities of practice: Learning, meaning, and identity*. New York: Cambridge University Press.

Wilensky, U. (1997). What is normal anyway?: Therapy for epistemological anxiety. *Educational Studies in Mathematics, 33,* 171–202.

Chapter 7

Modeling for Understanding in Science Education

Jim Stewart
University of Wisconsin–Madison

Cynthia Passmore
University of California–Davis

Jennifer Cartier
University of Pittsburgh

John Rudolph
University of Wisconsin–Madison

Sam Donovan
University of Pittsburgh

National science education reforms identify "understanding for all" as a central goal of school science curricula (National Research Council, 1996). Although the reforms established lists of what the scientifically literate student should know, translating these lists into curricula that foster understanding is not trivial. One reason for this difficulty is that the standards call for students to do more than memorize the conclusions of scientific research—they recommend that students also understand how scientific knowledge is constructed. We have found that a fruitful place to begin to realize this vision of scientific literacy is by developing curricula that are consistent with what scientific communities count as understanding. There is a growing consensus among educators of the value of engaging students in inquiry as part of their education in science. For us, scientific inquiry is a sense-making endeavor—a process of developing explanations about the natural world through participation in the complex activities of practice. Scientific knowledge is inextricably bound up with its generation and use, and to engage in scientific practice is both to demonstrate and to achieve understanding. We suggest that for stu-

159

dents, as well as scientists, understanding is the ability to function within a particular practice.

The inquiries conducted within scientific communities involve the development, use, revision, and assessment of causal models and related explanations. The significance of causal models is that they serve both as the mediating conceptual elements between the scientific community and the natural world and as the conceptual tools that research communities use to frame questions and provide the criteria by which *explanations* are judged. Models, however, do more than serve as templates for the conduct of empirical research—they can also become the focus of inquiry if they are thought to be inconsistent with empirical data, have internal inconsistencies, or are inconsistent with other established models. Science curricula should provide students with opportunities to use models to generate and interpret data, revise those models in response to anomalous data, and assess models for both empirical and conceptual consistency.

Students also need to have opportunities to participate in scientific argumentation. Specifically, we might expect high school students to engage in argumentation around developing explanations that link causal models and data, revising and assessing the adequacy of causal models and explanations, using models to pose questions and to design inquiries to answer those questions, and considering what standards or norms are to be used to assess the quality of claims.

In this chapter, we describe the curricula we created in collaboration with teachers involved in our multiyear professional development project and discuss the achievement in each content area of students using these curricula. Our Earth-Moon-Sun (EMS) astronomy, genetics, and evolutionary biology units involved students in developing, using, revising, and assessing causal models and related explanations—doing inquiry as scientists might and using discipline-specific forms of argument to support the explanations and models they developed in the course of their inquiries. If the goal of "understanding for all" is to be achieved, students must have opportunities to participate in realistic practice.

THE MUSE CURRICULA

Earth-Moon-Sun (EMS) Astronomy

The ninth-grade EMS unit begins with a multiday black box activity that focuses on developing criteria for judging models: explanatory adequacy, consistency with other models, and predictive power. In order to introduce students to a way of inquiring about and discussing the Earth, Moon, and Sun and their motions and to establish important class norms such as the need to support claims with evidence, teachers begin a discussion of the familiar phenomenon of day and night on Earth.

Instruction begins with a thorough introduction to the relevant EMS phenomena and appropriate data. During this time, students are asked to explore data, look for patterns, and develop questions about a phenomenon. Once a question is developed, students spend time in groups developing a causal model (or adding to an existing model) in order to explain the phenomenon. Students decide which celestial bodies are relevant to the data explored and how their motions might be used to account for the identified pattern. Once they work through the problem in small groups, the whole class comes back together to discuss ideas. The students use the criteria established during the black box activity to assess the ideas of their peers, determining whether each proposed model (a) can fully account for the data, (b) is consistent with their prior knowledge about the physical world, and (c) can be used to predict the results of new inquiries. As these discussions progress, students come to a consensus about the model components that can be used to explain the phenomenon in question. The primacy of group interactions throughout the unit provides opportunities for students to visualize, demonstrate, and represent phenomena and components of the celestial motion model as they work to ascertain a match between their own explanations and the observed data patterns.

In order to give a "complete" explanation of an EMS phenomenon, students have to put the relevant elements together into phenomena-object-motion (POM) charts (see Fig. 7.1), which include an explanation using both text and diagrams, and articulate the relationship between their celestial motion model (CMM) and the phenomenon in question (often using props such as inflatable globes, Styrofoam balls, and light sources).

This pattern of instruction is followed for the seven unit phenomena: day and night, sunrise and sunset, moonrise and moonset, the never-changing appearance of the Moon as seen from the Earth, Moon phases, seasons, and eclipses. As a culminating task, students work in groups to develop explanations for novel scenarios. Challenge problems such as the following require students to use their understanding of the CMM to explain novel phenomena from the EMS system or to apply their understanding of this model to a different system:

- If you were standing on the Moon looking at the Earth, describe how it would look throughout the Moon's orbit. Would it go through phases? How do Earth phases from the Moon relate to Moon phases from the Earth? (new phenomena to explain)
- As Venus is observed over a long period, it appears to go through phases. Why do we see phases of Venus? Why does the apparent size of

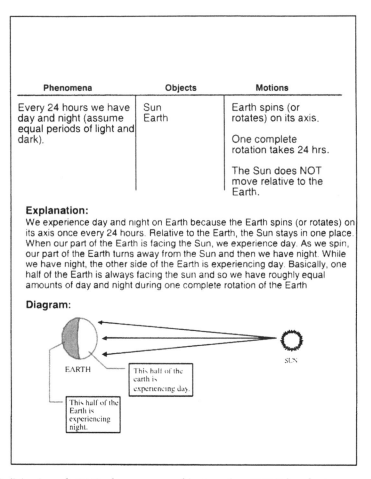

Phenomena	Objects	Motions
Every 24 hours we have day and night (assume equal periods of light and dark).	Sun Earth	Earth spins (or rotates) on its axis. One complete rotation takes 24 hrs. The Sun does NOT move relative to the Earth.

Explanation:
We experience day and night on Earth because the Earth spins (or rotates) on its axis once every 24 hours. Relative to the Earth, the Sun stays in one place. When our part of the Earth is facing the Sun, we experience day. As we spin, our part of the Earth turns away from the Sun and then we have night. While we have night, the other side of the Earth is experiencing day. Basically, one half of the Earth is always facing the sun and so we have roughly equal amounts of day and night during one complete rotation of the Earth

Diagram:

FIG. 7.1   Sample EMS phenomenon-object-motion (POM) handout.

Venus change throughout the cycle of phases? (application of CMM to different system)

Genetics

This 10th-grade, 9-week genetics unit focuses on the role of models. Causal models are introduced through students' reflections on their previous science experiences (in our studies these included their work in the EMS unit in the previous year). This introduction sets the stage for student inquiries into meiosis and classical genetics.

Meiosis is introduced through the use of family photographs of the teachers and their families. Students are asked to account for both similarity and individual differences within a family. Students are then given two sets of parental chromosomes and asked to "make a baby." As students attempt different combinations of parental chromosomes, the teacher gives them feedback, reminding them that the offspring need to be viable: for each pair of "baby" chromosomes, one has to come from the father and one from the mother. Students use this rudimentary meiotic model to discuss the genetic information inherited by the embryo, which results in the similarities to and dissimilarities from each other and their parents.

The section on classical genetics begins with a visit from "Gregor Mendel" (the teachers dressed up as the famous monk), during which time the students develop a model of simple dominance, which they then apply to pedigree data and cross data generated by a computer simulation (Genetics Construction Kit or GCK; Calley & Jungck, 2001). As part of this instruction, the students and teacher develop a representation of the simple dominance model that includes a notational system for identifying alleles (see Fig. 7.2). Once students are comfortable with the simple dominance model, the teacher introduces family pedigrees inconsistent with it. In the new pedigrees, there are three variations instead of the two they have seen with simple dominance. Students then spend several days revising their simple dominance model and testing their new models using GCK, eventually developing a codominance model.

The students are then confronted with anomalous data that cannot be accounted for with either model. At this point, the class is split into two groups. One group has the task of revising their model to explain Duchenne's muscular dystrophy (an X-linked disorder). The other half are given blood types of an extended family and asked to revise their models of simple dominance and codominance to account for four blood types. After working several days, the students present their revised models to the other half of the class. By the end of the unit, students will have developed and used four models: simple dominance, codominance, sex-linkage, and multiple alleles.

Evolutionary Biology

This 9-week high school course in evolution for juniors and seniors is designed to help students understand the central ideas in evolutionary biology, particularly Darwin's model of natural selection. The unit is intended to initiate students into the reasoning patterns of the discipline by engaging them in inquiry that requires them to develop, use, and extend Darwin's model of natural selection. Many evolutionary phenomena are not easily included in a high school course, and we found invaluable the use of case ma-

Mendel's Model of Simple Dominance

Variations   A and B

Alleles   1 and 2

Relationship between genotypes and phenotypes:

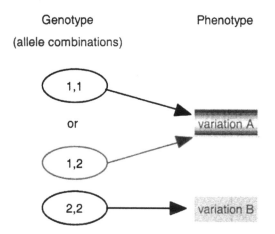

Genotype   Phenotype

(allele combinations)

Mendel also proposed that the "factors" or alleles of the parental plants segregated randomly during formation of offspring and that the alleles for one trait assorted independently from alleles of other traits.

Meiotic processes:

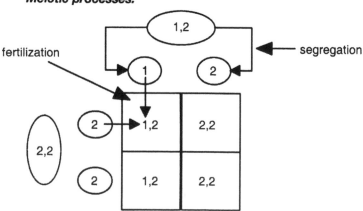

FIG. 7.2   Classroom representation of Mendel's model of simple dominance.

terials (e.g., readings and data that could be used to build explanations and to focus student discussions). Through the cases, students are involved in extended explorations, each designed to promote understanding of particular aspects of evolutionary biology. The first days of the unit are dedicated to establishing norms of classroom discourse in order to build a student research community that can work together in collaborative, yet critical, ways. This setup provides the students with a common language and analytical tools with which to assess each other's arguments.

Following this introductory activity, students explore the assumptions of the models proposed by Paley, Lamarck, and Darwin, which each, in its own way, accounts for the Earth's species diversity. In the process of examining these models, students compare and contrast Darwinian and non-Darwinian assumptions. Following this introduction to the natural selection model and the comparison of the underlying assumptions of the three models, students are given the task of developing a Darwinian explanation for a simple adaptation. On a general level, the development of a Darwinian explanation is straightforward—students are expected to use their understanding of Darwin's model of natural selection (including variation, superfecundity, selection, and heritability) to explain the origin and maintenance of adaptations. The form that such an explanation takes requires students to weave into a historical narrative the descriptions of the variation within the population, the selective advantage of certain variations when compared to others, and the role of inheritance. In constructing Darwinian explanations, students need to explain the change that has occurred over a period of time in the frequency of a particular trait and indicate the historical contingencies on which their explanation is based.

Once students have initial experience composing Darwinian explanations, they are given a data-rich case from which they are expected to develop a Darwinian explanation. Their task is to work in research groups to create an explanation of the change in one of two seed-coat characters over time (using the natural selection model) and tie data from the case materials directly to each component of the model. The culminating activity of the case is a poster session during which each research group presents their poster to other research groups and spends time interacting with these other groups about the adequacy of explanations.

The final two cases provide additional opportunities for students to interact about evolutionary phenomena. In the first of these cases, students explain the similarity in the color patterns of viceroy and monarch butterflies. Applying the model of natural selection to this less-than-straightforward case necessitates a rich understanding of the model and the assumptions on which it is based. Once the students finish writing their explanations for the monarch–viceroy case, they come together to read and critique explanations. The class then discusses the case generally, describing the overall arguments and citing specific pieces of data to support their claims. The students then go on to discuss the origin of bright coloration, which normally leads to other interesting questions that resemble questions that evolutionary biologists might ask.

In the second case, students develop a Darwinian explanation for the difference in coloration between male and female ring-necked pheasants. Once they formulate an explanation supported by evidence drawn from the case materials, they develop a research question that allows them to investigate some component of their explanation and present their explanations and research questions to the class in a "competition" for research funds. Once all groups have presented, the students discuss the merits and shortcomings of each proposal and then decide which proposal to fund.

FINDINGS FROM OUR RESEARCH ON STUDENT UNDERSTANDING AND REASONING

Earth-Moon-Sun (EMS) Astronomy

The findings reported here are based on a study of the entire ninth-grade class (10 course sections, 220 students) in a suburban school. One section (22 students) was designated as a *focus classroom*, and 13 students (7 from the focus classroom and 6 from two other sections) were designated as *focus students*, for a total of 28 *treatment students* and two *treatment teachers*. In the focus classroom, we collected daily field notes and audiotapes; at least two days a week we also collected field notes in other classrooms taught by each of the four teachers. In addition, we conducted a series of five 15- to 40-minute interviews with the focus students and collected all written work (exams, quizzes, notebooks, projects) done by the treatment students. We also collected the final exams and posters from a subset of students in each of the 10 sections.

Prior to the introduction of each new topic, students completed a short pretest. During the unit, the students took three quizzes designed to assess their understanding of and ability to apply their celestial motion model (CMM) to explain or predict novel situations. At the end of the unit, students were given back their pretests (unmarked by the teacher) and asked to review and correct their answers in preparation for the comprehensive exam, which included items designed to assess EMS and inquiry learning outcomes.

Based on their performance on a pretest and posttest, it was evident that students made significant gains in their understanding of EMS learning outcomes. Figure 7.3 summarizes the percentage of *treatment* students able to give complete and correct explanations for the indicated phenomena before and after instruction. Complete explanations are those in which the

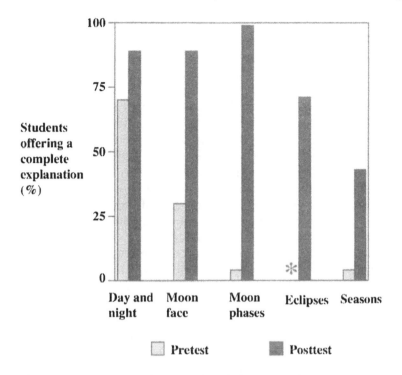

FIG. 7.3 Comparison of student EMS knowledge pre- and postinstruction.
*Pretest data on student knowledge of eclipses omitted because students were not
specifically asked to account for the phenomenon on the pretest.

relevant objects and motions within the CMM are identified and the causal
relationship between these and the relevant phenomenon is explicitly
stated. Students' written work on quizzes and exams indicated that their understanding of the underlying causes of each phenomenon increased post
instruction (see Fig. 7.3).

One issue related to the pre- and posttest data bears explanation. On the
pretest, students were asked to identify which of four diagrams represented
the appropriate alignment of the Moon, Earth, and Sun during a solar and
then a lunar eclipse and to describe what caused the eclipses in question. In
response to this pretest question, 58% of the students were able to correctly
identify the representation of at least one type of eclipse and describe how
the eclipse was caused by one celestial body blocking another from view (in
the case of a solar eclipse) or one body casting a shadow on another (in the
case of a lunar eclipse). All 28 treatment students were able to complete this

task on the final exam. On the exam, we asked the additional question, "Why don't we have a solar and lunar eclipse every month?" In response to this question, 71% of the students were able to describe the way in which the Moon's orbit intersects the plane of Earth's solar orbit during the full and new Moon phases only twice in a 12-month period. Such an explanation constitutes a correct and complete answer for this question. On the pretest, none of the students had described the Moon's tilted orbit or discussed the importance of the intersection of the Moon's orbit with Earth's solar orbit during certain Moon phases to cause eclipses. Because we didn't pointedly ask them to account for the 6-month data pattern, however, we can't be certain that they wouldn't have mentioned these causal relationships if given the chance. Consequently, the summary of students' pretest data for eclipses is omitted from Fig. 7.3. We note, however, that it seems unlikely that many of the students would have known its cause: On the pretest, only one student was aware that eclipses occur twice a year (the majority of students believed that they occurred every 12 months).

The majority of the students learned the science concepts in the EMS unit. An examination of students' written work and transcripts from interviews with a subset of those students suggested that they developed an understanding of the EMS concepts through multiple opportunities to represent their ideas and use the CMM to account for various data patterns. Additionally, the students demonstrated that, over time, they were increasingly able to reason from effects to causes and to isolate specific causal relationships. In other words, in addition to learning EMS concepts, the students' reasoning changed.

In order to understand better how the EMS unit supported these types of student achievement, it is useful to look more closely at how their understanding developed. Thus, we have selected a representative student "Curtis," whose work is discussed next.

Curtis, like many of his peers began the unit with considerable "final form" CMM knowledge. Before the unit, Curtis knew that the Earth spins on its axis, the Moon orbits the Earth, the Earth orbits the Sun, and the Earth is tilted on its axis, but he was unable to account for the phases of the Moon or the seasons. He was initially able to explain the cause of day and night by pointing to a specific motion, the spin of the Earth, but was unaware, for example, that the same side of the Moon always faces Earth; that the Moon is sometimes visible in the daytime; that approximately 28 days pass from one full Moon to another; and that the Northern Hemisphere actually experiences winter when it is closest to the Sun. He was also unfamiliar with some of the *causes* of these phenomena.

For Curtis, participating in inquiry was an important first step: Once he learned about the phenomena, opportunities to identify and articulate cause-and-effect relationships helped him to develop understanding of the

EMS target concepts. One unfamiliar phenomenon, the unchanging Moon face, was particularly difficult for him, and he clearly struggled to isolate its cause. In a POM chart, Curtis correctly identified the objects involved (the Earth and the Moon) and the two relevant motions (the moon's revolution around the Earth and the Moon rotation). He knew that the relevant motions are the Moon's rotation and revolution, but he was not yet able to use this knowledge to specifically account for the phenomenon. Instead, Curtis offered a number of previously learned facts related to the Earth (such as its tilt) in an attempt to cobble together an acceptable explanation.

As the unit progressed, Curtis became much more skilled at isolating particular causal relationships and reasoning both from cause to effect (such as when answering extension questions) and from effects to causes (such as when determining the reason for Earth's seasons at the end of the unit). On his final exam, he elaborated on his explanation of Moon phases, adding a new piece of information and a very specific motion:

> This [regular order of Moon phases seen from Earth] occurs because the sun only illuminates half of the moon at one time. So while the moon is orbiting the earth we are only able to see a certain amount of the moon at one time.

His answer was accompanied by the representation shown in Fig. 7.4. Curtis also clarified that in order to explain the specific order of the Moon's phases, it was important to note that the Moon orbits the Earth in a counter-clockwise direction as viewed from the North Pole.

On his quizzes and final exam, Curtis correctly and completely explained the cause of each of the phenomena studied during the EMS unit, demonstrating that he had developed a solid understanding of the target science concepts through inquiry within a scientific community.

A comparison of Curtis' pretest and exam scores demonstrates that he achieved many of the EMS learning outcomes. On the pretest, Curtis answered only 11 of the 21 questions correctly. He began the EMS unit familiar with most of the phenomena and had knowledge about the motions associated with the Earth, Moon, and Sun, but he was frequently unfamiliar with the patterns associated with phenomena and generally unable to connect what he knew about celestial motions with specific phenomena in order to explain them. In contrast, he earned a 99% on his comprehensive final exam. By the end of the unit, Curtis, like many of his peers had become skilled at identifying causal relationships and explaining phenomena using the CMM.

Contrasting students' work before and after EMS instruction, what stands out most is not the difference in *what* they knew, however, but in *how* they knew it. The majority of the focus students began the unit knowing many facts related to the CMM. However, very few of them were able to use

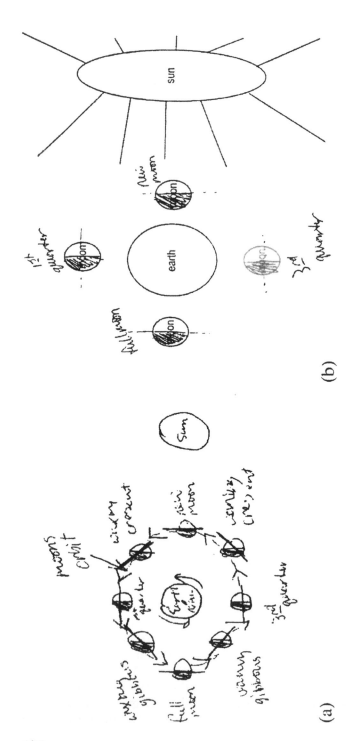

(a)

(b)

FIG. 7.4 Curtis' representations of Moon phases. (a) This drawing accompanied his written explanation of phases on Quiz 3 (October 22, 1999). Note that Curtis drew and labeled eight Moon phases. The drawing is from an "outer space" perspective: The entire Moon is shown, and the side facing the Sun is illuminated. The lines dissecting the Moon are drawn to show which part is visible from Earth. (b) This drawing accompanied Curtis' final exam explanation of phases (November 9, 1999). Curtis completed this drawing to show the portion of the Moon lit by the Sun and used dissecting lines to indicate what part of the Moon is visible from Earth in each of the four phases shown.

their knowledge to explain the direction of Sun rise and set, the phases of the Moon, or seasonal temperature fluctuations. Their work within the EMS unit, however, helped them to connect this knowledge to concrete experiences and to transform disjointed facts into a dynamic celestial motion model with real power to explain and predict phenomena.

The EMS unit helped students develop the ability to use the CMM to account for phenomena in two important ways: First, there were varied opportunities to represent and visualize phenomena and components of the model. Representation enabled students to identify elements of the CMM and associate them with particular phenomena.

The second way in which the unit supported students' developing dynamic knowledge was through the adoption of a particular, explicit set of criteria for assessing and justifying explanations. To be satisfactory, an explanation had to be able to account for data and be consistent with relevant prior knowledge. Additionally, students learned to test their own understanding by making predictions with their models while solving extension questions and challenge questions. It was through this flexible use of their celestial motion models that students like Curtis became most confident about their understanding.

Genetics

The research presented here extends earlier studies (see Cartier & Stewart, 2000; Finkel & Stewart, 1994; Hafner & Stewart, 1995; Johnson & Stewart, 2002; Wynne, Stewart, & Passmore, 2001) to describe the longevity of students' understanding and reasoning abilities. Because we were interested in whether the ability to reason genetically, once established, persisted over time, we chose seven students who were reasonably successful in response to instruction early in the class. These students were interviewed four times during the unit to provide baseline data on their understanding of the genetics. One year later, the students were interviewed again in order to determine their understanding of meiosis, simple dominance, codominance, multiple alleles, and X-linkage; and to explore how they used this understanding to explain novel inheritance phenomena. Our report here is based on the interviews conducted one year after instruction.

Knowledge About Meiosis. In the interviews, students assigned genotypes to a pedigree in response to questioning, frequently pausing to construct Punnett squares. When asked what the Punnett square represented or why they assigned each person two alleles, their responses demonstrated that they understood the details of meiosis, including the processes of segregation and independent assortment. When presented with an albinism pedigree (a novel scenario; see Fig. 7.5), all seven students immediately identified it as a trait governed by simple dominance inheritance, with the albino variation being recessive to normal pigmentation. Thus, when assigning genotypes to albino individuals on the pedigree, all of the students labeled them as 2,2 (homozygous recessive). As a challenge question, students were then asked to explain an unusual situation: a heterozygous nonalbino father and a homozygous nonalbino mother producing an albino daughter. Let's look in depth at one student's work.

Matt had earlier identified the parents of this albino daughter as heterozygotes (1,2), reasoning (reasonably) that each parent had contributed one allele for albinism (a 2) to their daughter. After being told that in fact the mother was homozygous dominant (1,1), but the daughter was somehow albino (2,2; homozygous recessive), he was asked to theorize what could have happened:

Matt: It could be that in the pairing of [pause], the daughter is still a
             2,2? ... Or just a 2?
I: She is a 2,2.
Matt: Okay. Hm.
I: Will you finish the thought that you have been having so I can
             hear what it was?

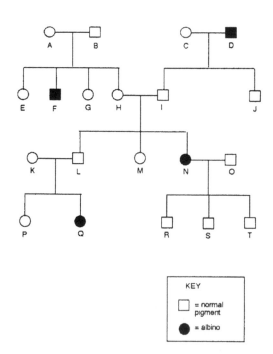

FIG. 7.5   Albinism pedigree.

Matt: Um. It could be that in the pairing, it could be in the pairing of the chromosomes that the chromosome that had this trait on it did not jump into the, didn't separate appropriately.... So one egg cell had two 1s in it, and the other had none at all.... And if this is a 2,2 [pointing to the daughter], it would also have to be an extreme case in which the two 2s or, no, there wouldn't be two 2s. Because one would be a 1 and the other would be a 2. So, the 2 would be passed on to the daughter [from the father]. And the daughter wouldn't get a 1 or a 1 from the mother. So the daughter would only have one [a single] 2 of that chromosome. And it would not have a second chromosome controlling that trait.

In response to the anomaly (an albino daughter from these two parents), Matt had immediately begun to think about what might have happened during meiosis, wondering first if possibly the daughter had only one allele, the recessive 2 (from her father). After being told that the daughter had the two alleles (2,2), he began to consider what could have happened with the father's chromosomes during meiosis, but seemed caught by the fact that the father had only one 2 to give. Thus, Matt ended by saying the daughter would only have one recessive allele and "it would not have a second chromosome controlling that trait." When told that the daughter had in fact received two copies of that chromosome from her father, Matt continued:

If in the process in which the original cell from the father copies itself before splitting in two, in half, if in that process of copying those two [the 1 alleles] didn't jump into the same cell when splitting, so you had the 1 and the 2, when they copy, you get another 1 and another 2, and let's see. It would be another 1 and another 2. And they would then split like that [unevenly so three chromosomes went to one cell and one to the other]. If in the splitting process they split like that, where the 2s stayed together and had a 1, so she had, the father cell had three of the chromosome ... then when this [the new cell with three chromosomes] split in half it split down the middle. Two would still be, the two 2s would still be together. The 1 would go off on its way.... Of the four that are produced, two of them have nonalbinism. One could care less because there is nothing controlling albinism. And the fourth has two traits for albinism.

Matt's reasoning around this question provided evidence that he understood the meiotic model. He was able to use his understanding of basic meiotic processes to reason about a novel question and could imagine how a disjunction error during the meiotic division could result in a child who did not receive any information for a particular trait from one parent and "in an extreme case" received two pieces of information from the other parent. He later went on to say that although his explanation made sense to him, he thought the whole occurrence was extremely unlikely.

Throughout the interviews, all of the students addressed the question of how a homozygous dominant mother could give birth to a homozygous recessive daughter. This question revealed a great deal about their understanding of meiosis. All of the students identified a mistake in meiosis as a possible cause. Four of the students postulated a mistake in the formation of the sperm in which the doubled chromosomes "didn't split up all the way." Only two students, however, proposed a similar mistake in the mother's egg. The other two students suggested that the presence of two recessive alleles in the sperm could have "overpowered" the mother's allele. The last two students were able to describe nondisjunction events only after they were told explicitly to think about meiosis. It is noteworthy, however, that even students who did not raise the possibility of meiotic errors themselves still demonstrated an understanding of how such errors could account for this phenomenon.

All of the interview participants were also able to describe the processes of segregation and independent assortment when given an opportunity to do so, provide a general description of gamete formation, and reason about the nondisjunction question (see Table 7.1 for a summary).

Knowledge of Meiosis Displayed When Explaining Simple Dominance. All of the interview participants were given a simple dominance pedigree (albinism) to explain. All of the students identified albinism as a recessive variation. They used two basic strategies to do this. At times, they were able to identify simple dominance based on a recognition of empirical patterns

TABLE 7.1
Student Retention of Knowledge of Meiosis

Concept	Erin	Tony	Tim	Mike	Karen	Paul	Matt
Meiosis							
description of process	K	P	K	P	P	K	K
explanation of two alleles per individual	K	K	K	K	K	K	K
segregation	K	K	K	K	K	K	K
independent assortment	K	ND	K	K	P	ND	K
fertilization	K	K	K	K	K	K	K
chromosome #	I	I	K	K	K	K	I
nondisjunction event	P	P	P	P	P	P	K

Note. I = incorrect; K = demonstrated knowledge; ND = no data; P = partially correct.

alone, typically using a cross such as K and L because they knew that when two unaffected individuals have an affected child, the parents must be het-erozygotes. This type of response required only that students remembered the empirical pattern associated with recessive inheritance.

Although many of the students used empirical pattern recognition at some point in their discussions of the albinism pedigree, they did not rely solely on this strategy. All of the interview participants used their under-standing of meiosis to assign genotypes to the individuals in the pedigree. They reasoned with the meiotic model in two ways. First, they used their un-derstanding of meiotic processes to determine the genotypes of unknown individuals based on the genotypes of their parents or offspring. From this section of the interview, it was clear that all of the students drew on their un-derstanding of meiosis to assign genotypes to individuals in the simple dominance pedigree. Although it was possible to reason about genotype as-signments using only empirical pattern recognition, none of the students relied solely on this strategy. Instead, they invoked meiosis to reason about allele distribution in the family. Taken together, the evidence from the final interviews one year post instruction point to students' recognition of a clear connection between meiosis and simple dominance.

Knowledge of Classical Genetics Models: Simple Dominance, Codomin-ance, X-Linkage, and Multiple Alleles.   As previously noted, the students were able to successfully assign genotypes to phenotypes on a simple dom-inance pedigree. There was evidence in all of the interviews that students remembered a great deal about the simple dominance model of inheri-tance. Similarly, all of the interview participants successfully used a co-dominance model to explain a pedigree they were given displaying the trait of hair texture.

For the multiple-alleles and X-linkage models, students were not asked to reason with pedigree data, but were simply asked to recall the basic ele-ments of these complex inheritance patterns. During instruction they de-veloped a multiple-alleles model to explain the inheritance of blood types in humans. During the interviews, all students remembered that there were three alleles instead of two, and five students explicitly discussed how there were differing allelic interactions at work in the blood types model. Four of the students also provided a complete description of the X-link-age model and discussed the X-linked traits, noting that the male has only one allele and explaining why having only one allele would alter the inher-itance pattern (see Table 7.2 for a summary of student knowledge of all ge-netics models).

Again, let's look in depth at students' interviews in regard to their under-standing of multiple alleles and X-linkage, beginning with multiple alleles. During initial instruction, students had been presented with a family his-

TABLE 7.2
Student Retention of Knowledge of Classical Genetics Models

Concept	Erin	Tony	Jim	Mike	Karen	Paul	Matt
Simple dominance							
assigned 2 alleles/individual	K	K	K	K	K	K	K
identified 2 alleles/population	K	K	K	K	K	K	K
identified 3 genotypes and 2 phenotypes	K	K	K	K	K	K	K
used empirical pattern recognition strategy	K	K	K	K	K	K	K
Codominance							
assigned 2 alleles/individual	K	K	K	K	K	K	K
identified 2 alleles/population	K	K	K	K	K	K	K
identified 3 genotypes 3 phenotypes	K	K	K	K	K	K	K
Multiple alleles: blood type							
identified 3 alleles/population	K	K	K	K	K	ND	K
described varying allelic interaction	K	K	ND	K	K	ND	K
identified 4 phenotypes	K	K	K	I	K	ND	K
X-linkage							
assigned 1 allele in males	K	F	K	K	ND	K	K
described altered allelic interactions	K	F	K	K	ND	ND	K

Note. I = incorrect; K = demonstrated knowledge; ND = no data; P = partially correct.

tory and pedigree that showed four distinct phenotypes controlled by three alleles. All of the students were able to recall this when asked. After this, they were asked if they could imagine a three-allele, three-variation model. "Tony's" reasoning was typical:

> Okay. So there is an A, a B, and an O trait. And what I thought it was, was O is a recessive trait ... and A and B are just that, they are like these two [referring to the codominance pedigree he had just explained]. Neither one of them are dominant. They are both, like, the same. So you have your A parents and your B people. You have your A people, your B people. And you have AB people, which are different. And then you can have AO people, which is, like, almost the same as this [pointing to A, A] ... They both show the same thing because O is recessive. So it [the A] takes it over. Then

you have a BO, which is the same as this [pointing to B, B] because the B is only showing. And the only way that an O blood type can be shown is if it is an OO ... There are three alleles, but four different ways people can look.

As he was talking, Tony wrote out the six genotypes that could be made from the combination of the three alleles. He also described how the interactions among alleles were sometimes codominant (in the case of the A and B) and sometimes dominant (as in both A and B are over O). In this example, he used his knowledge of allelic interactions as depicted in the simple dominance and codominance models to explain the genotype-to-phenotype mappings he had made.

When asked if he could use a three-allele model to explain three variations, he switched to X, Y, and Z to represent alleles. As he began to reason about this problem, Tony assigned two alleles per individual and made the assertion that the homozygotes (X,X; Y,Y; Z,Z) would have distinct phenotypes ("because they don't have anything in common at all with each other"). He then began to think about where the other possible genotypes would fit. Eventually he seemed to settle on an idea that "order matters." At this point, he went on to identify nine different genotypes based on the order of the alleles. He matched those with the corresponding first letter to either the X, Y, or Z phenotype and ended by saying it was "the only way I can think about this off the top of my head." When asked to think about this issue as if "order doesn't matter," he continued:

Well, like this one [Y, X] would be taken out because these two are the same [pointing to X, Y and Y, X]. And this one [Z, Y] would be taken out because these two are the same [Y, Z and Z, Y]. And this one [Z, X] would be taken out because these two are the same [X, Z and Z, X]. So you are back to this thing again [pointing to the ABO mapping he had done]. You have your three [homozygotes]. They are all the same. Then you have, they are mixed with each other, or if they are, you know, when they are all mixed with each other ... So I don't understand how you can make a model where there is only three traits because [long pause], what if let's say X was dominant? ... Then this [X, Y] would look like that [X, X], right? ... And that [X, Z] would look like that [X, X]. If it [X] is dominant over both of them. And Y is dominant over Z but not over X. Then you can only have three.

When asked if his explanation made sense, Tony replied:

Yeah. Because X is the most dominant trait. So anything that has X in it would show X. Y is below that. It is more dominant than Z but less dominant than X. So when Y and Z make Y showing, when Y and X mix, X is shown. And Z is, yeah, that makes sense.

In essence, Tony had mapped the six remaining genotypes onto the three phenotypes by imagining that X was the "the most dominant." Throughout his reasoning, Tony drew on the idea that more than one genotype could map onto a particular phenotype. This idea, that one allele can be dominant over another, is of course a key concept in the simple dominance model. Tony used his knowledge of dominance interactions among alleles to successfully develop a three-variation, three-allele model.

In summary, the interviews provided evidence that the students remembered a great deal of the ideas involved in the classical genetics model, especially in their attempts to develop a three-allele, three-variation model. In order to develop this model, the students had to consider a variety of possible allelic interactions. In doing so, they had to (a) draw on their knowledge of multiple alleles to imagine how three alleles could be present in a population while only two alleles are present in individuals; (b) realize, using their knowledge of simple dominance, that more than one genotype could correspond to a particular phenotype; and (c) simultaneously reject the possibility of codominant allelic interactions. Students' abilities to address this scenario demonstrated understanding of the range of classical genetics models as evidenced by their ability to think with the model ideas. Even a year after instruction, students demonstrated that they remembered a great deal of what they had learned and maintained an ability to reason with the models of meiosis, simple dominance, codominance, multiple alleles, and X-linkage.

Evolutionary Biology

In the findings presented here, we explore students' ability to extend what they had learned about natural selection to explain the changes in heritable characteristics of populations over successive generations. Eighteen students (11 females, 7 males) participated in the study. The discussion is based on student work on pre- and posttests, audiotape data from student classroom discussions, and interviews with individual students.

Student Reasoning About Selective Advantage.   In order for students to grasp what it meant for an organism to have an advantage in a Darwinian sense, they had to come to recognize the interplay between survival, reproduction, and inheritance. On the pretest, many students associated natural selection with survival, yet they did not recognize that simply surviving into old age is not sufficient in evolutionary terms. By the end of the course, however, they came to appreciate the connection between survival and reproduction and were able to discuss the importance of inheritance. On the posttest, students also demonstrated a deeper understanding of the mechanism of natural selection and how it acted in nature. Even when students

were unsure of the exact selective advantage conferred by a particular variation, they still posited that there must be one in terms of survival and reproduction. Although on the final exam several students were still vague in
their descriptions of the advantage of white fur for the polar bear, they did
incorporate discussion of selective advantage in terms of survival and reproduction in their answers. They seemed to operate under the assumption
that in order for a particular variation to become more common in a population over time, it had to confer a selective advantage on the *individual* that
possessed it. As a result, students sometimes struggled to explain the selective advantage of variations that initially did not seem to be advantageous.

The pheasant case, for example, required students to struggle with the
apparent disadvantage of bright color in the males. In the excerpt below,
"Mallory" was discussing her pheasant explanation during an interview
conducted as part of the research:

> We said that ... bright coloring came when it got to a mature age [inaudi-
> ble], so it was more noticeable for the females in the population to notice
> them, to find them to mate easier, but it was also disadvantageous to them
> because it made the predators, it made them easier for the predators to
> see, so a lot of them are killed, too. It just happened over time that the
> population just kind of got brighter and more noticeable for the females,
> maybe because the ones that weren't as bright never got to mate, so they
> never got to pass that on, and the ones that were brighter mated more of-
> ten and were able to pass that trait on to the male sons ... I don't know,
> maybe that helps explain why there's only a few males that populate a
> whole population, 'cause if a lot of them were killed off by the predators
> then there's only a few, and maybe they're smarter, and they hide from the
> predators. I don't know how it kept on coming about. I'm sure there's
> males that are camouflaged, but those ones don't get to mate though, so I
> guess that's the variation that's in the population. Some of them survive
> this season or whatever, and some of them don't.

This student identified both an advantage and a disadvantage for bright
plumage, but she made no attempt at first to reconcile these two views.
When pushed to do so, she thought aloud as she tried to come to a view that
made sense to her. By the end of this excerpt, she was comfortable with the
idea that an advantage in terms of reproduction might potentially outweigh
an advantage in terms of longevity.

Students also had interesting discussions during their work on the monarch–viceroy cases and in developing explanations for mimicry. During one
discussion, students came to realize that population-density data were useful in imagining how the predators could begin to associate bad taste with
bright coloration. Although students were concerned with the apparent disadvantage of bright coloration in terms of predation for individual mon-

archs, they could imagine how bright coloration might be advantageous to the population as a whole if predators began to associate brightness with "getting sick." This differentiation between an advantage for an individual organism and an advantage for the entire population is a key feature in evolutionary reasoning. Students also showed a sophisticated understanding of Darwinian explanations in dealing with a complex species interaction—the potential problem for the monarchs if they were outnumbered by the more palatable viceroy mimics.

Student Reasoning About Variation.   The theme of variation was woven throughout the instruction and highlighted as a key concept in the natural selection model. In our analysis, we looked closely at student descriptions of initial and final populations to determine how they thought about variability before and after a selection event. Prior to instruction, fewer than half the students even mentioned an initial population (none explicitly), and none of these described it as variable. The seven students who did mention an initial population were simply repeating the original prompt. Students who made a passing reference to the Albermarle tortoises (identified in the question) as ancestral to the more recent saddlebacks were coded as mentioning an initial population, but in no response was there a reference to variation within that population. In fact, most of the students simply described the advantage of the derived trait for an individual and were not attempting to explain how this change occurred in a population of organisms over time (despite being asked to do so in the question). In contrast, *all* students described variation in the initial population on the posttest, even though the prompt on the final exam contained no mention of variation. Students had adopted a Darwinian perspective that allowed them to account for a shift in the frequency of particular variations within a population over time rather than simply describing the apparent functional advantage of a present-day trait. Not only did students describe a shift from one population to another, they also linked the two by describing the advantage of the trait in an evolutionary sense.

Let's look at an example where students were attempting to understand how bright coloration might have come about in a population of monarchs. One student considered how the variation might have looked in the past in terms of the population as a whole, and students went on to talk about the possibility that initially there could have been multiple butterflies "brighter" in color in a population. They asserted that even if the trait of brightness originated with one pair of butterflies, "butterflies have a lot of offspring," and potentially several brighter butterflies could hatch from a given batch of eggs, giving rise to a variable population. Note that these students were thinking of at least two phenotypes in the ancestral population—they did not seem to realize that the populations might have many slight variations.

These discussions about initial variability in the population led students to further consider the ancestral population and question ways to define the starting point of evolutionary change:

Anna: I feel like we're saying that everything had to evolve from a dull color. What if it originally was bright? That's just how it was. Just like every species has variation, that was just that's what it was. It was bright for no particular reason.

Mark: But it had to evolve from something.

After a brief discussion of other animals and what they knew about ancestral traits, Mark brought the conversation back to defining starting point:

Mark: But it [brightness] had to evolve from something sometime?
             ...
Rob: Do you think monarchs were always poisonous?
Mark: As long as they've been eating milkweed.
Rob: You think they always laid their eggs on milkweed?
Tim: They had to develop the ability to eat poisons without dying and pass on that trait to their offspring.
Doug: Well, how do you define a monarch? That's the question.

The class became animated as they debated how to define a population in the past given information only about the present. Once students were able to engage with the idea of a variable population, they were able to have substantive conversations about the mechanism that brought about change from that initial state. Their focus on population-level questions shows that they understood that natural selection, acting on individuals, could result in changes to the population.

Changes in Students' Articulation of Evolutionary Concepts.   In our analysis of students' conceptualization of evolution, we noted a change in the language students used in their explanations. In their pretest responses, students used language that implied a view of evolutionary change as absolute, automatic, or deterministic. Even when they discussed the ideas of survival or reproduction, they did so by saying "they will survive," "they all reproduced," and so on, implying a view of natural selection as an event easily predicted. In contrast, on the responses to posttest questions, students used what we have come to term "probabilistic language" in reference to differential survival, differential reproduction, and the likelihood of inheritance. After instruction, students often qualified statements about survival, reproduction, or inheritance with phrases like "they had a better chance" or "they were more likely to reproduce"—indicating an increased understanding of the uncertainty inherent in natural selection. This growing sensitivity to the language in narrative explanations, taken together with students' use of words in relative descriptions and their explicit posttest declarations about variability (absent in pretests), indicates to us that students had begun to think in terms of a continuum rather than in absolutes and provides compelling evidence that the students came to an increased appreciation of the complexities of evolution by natural selection.

FINAL COMMENTS

Prior to instruction, many of the students displayed the misconceptions in EMS astronomy, genetics, and evolution commonly reported in the literature. By the end of these courses, students were able to weave their understanding of individual concepts together to develop explanations for complex phenomena. The findings that we report here are significant.

We believe these courses provide one vision of the ways that reform recommendations can be translated into classroom practice. Developing curricula that allow students to learn through inquiry requires a commitment to providing instruction true to the particular scientific discipline and to creating opportunities for students to *use* the central models in that discipline to account for relevant phenomena. Too seldom do science classrooms reflect the intellectual activity of science, and all too often students come away believing that having an aptitude for science means having a strong ability to memorize. Yet, as we have shown here, when students are given time and classroom environments that encourage inquiry, they are capable of sophisticated reasoning.

At the outset of the EMS unit, many students were unaware of patterns associated with natural phenomena, and students often held misconceptions. By using an inquiry framework to organize the curriculum and structure student work, we were able to introduce each new topic by exploring relevant data, with an initial focus on identification of patterns occurring in the natural world. Students were offered varied opportunities to *represent* and visualize phenomena and components of the model. They adopted a particular, explicit set of criteria for assessing and justifying *arguments*—and they learned to test their own understanding by making predictions with their models while solving extension and challenge questions. As a direct result of our curriculum, we noted significant gains in students' awareness of seasonal trends, Moon phases, and the frequency of eclipses as well as in their ability to explicitly account for celestial motion patterns. Contrasting students' work before and after instruction, what stands out most, however, is not the difference in *what* they knew, but in *how* they knew it. The majority of students began the unit knowing facts related to celestial motion, but this initial knowledge was "static"—very few were able to use what they knew to explain natural phenomena. Their work within the EMS unit helped them to connect their knowledge to concrete experiences and to transform disjointed facts into a dynamic celestial motion model with real power to explain and predict phenomena.

Overall, students in our genetics course developed a more mature scientific epistemology, shifting from an expectation of scientific *proof* to one of explanatory power or data–model *fit*. Initially, students described the ways in which *data* were brought to bear *in proving theories*, but at the end of the course, and perhaps due to their own experiences with modeling, students described ways in which *models* were brought to bear *in making sense of data*. This shift away from seeking "proof" in science toward offering empirical support for scientific models was never explicitly addressed in class, but seems instead to have been a natural consequence of the students' own experiences revising and justifying genetic models. But students did not just learn about the nature of scientific models. *Through model-based inquiry*, students not only came to a rich understanding of genetics, both of the central concepts of transmission genetics and of important epistemological aspects of genetic practice, but, as importantly, they developed a meaningful understanding of the nature of scientific practice.

Students in our evolutionary biology course developed sophisticated understanding that included, in addition to the ability to identify and discuss concepts relating to evolutionary change, the ability to use those ideas to formulate explanations appropriate to the discipline. Specifically, students became adept at using the natural selection model to account for realistic data and to develop Darwinian explanations for that data. As a result of their experiences, the students were able to reason about evolutionary concepts such as variation and differential survival as well as to use those concepts to explain changes in populations over time. Our data analysis indicates that students developed a rich understanding of the natural selection model and used that understanding to reason in discipline specific ways about evolutionary phenomena.

What, finally, do we believe is gained by teaching *through* inquiry? Given the success of our students in this inquiry learning environment, structured integrally into all three courses, we propose that instruction that affords students opportunities to participate in authentic inquiry, data collection, and analysis as well as reasoning and argumentation about explanatory models will support enhanced student understanding and strong levels of student achievement. We also suggest that only in this context do students acquire dynamic knowledge—*knowledge that can be applied* in making sense of the world. "Knowing" components of a conceptual model without having a sense of the phenomena that the model explains or the evidence on which the model is based is shallow understanding at best. Our data demonstrate that students can and do develop dynamic understanding of scientific concepts through participation in inquiry-based curricula.

REFERENCES

Calley, J., & Jungck, J. R. (2001). Genetics construction kit (The BioQUEST Library VI) [Computer software]. New York: Harcourt Academic Press.

Cartier, J., & Stewart, J. (2000). Teaching the nature of inquiry: Further developments in a high school genetics curriculum. *Science and Education, 9,* 247–267.

Finkel, E., & Stewart, J. (1994). Strategies for model-revision in a high school genetics classroom. *Mind, Culture, and Activity, 1*(3), 168–195.

Hafner, R., & Stewart, J. (1995). Revising explanatory models to accommodate genetic phenomena: Problem solving in the context of discovery. *Science Education, 79*(2), 111–146.

Johnson, S., & Stewart, J. (2002). Revising and assessing explanatory models in a high school genetics class: A comparison of unsuccessful and successful performance. *Science Education, 86,* 463–480.

National Research Council. (1996). *National science education standards.* Washington DC: National Academy Press.

Wynne, C., Stewart, J., & Passmore, C. (2001). High school students' use of meiosis when solving genetics problems. *The International Journal of Science Education, 23*(5), 501–515.

Chapter 8

Learning Mathematics in High School: Symbolic Places and Family Resemblances

Ricardo Nemirovsky
Apolinario Barros
Tracy Noble
Marty Schnepp
Jesse Solomon
TERC

The Urban Math project centered on classroom teaching experiments in three high school math classes: a second-year integrated math class taught by Jesse Solomon at the City on a Hill public charter school in Boston, MA; an AP calculus class taught by Marty Schnepp at the Holt High School in Holt, MI; and a bilingual Algebra I class taught by Apolinario Barros at the Jeremiah Burke High School in Dorchester, MA. Over several years, these classrooms have fostered outstanding student learning by both traditional (e.g., grades, college admittance) and nontraditional (e.g., student engagement, class participation) criteria. Through the teaching experiments, we strived to identify key aspects of the teaching practices and instructional design that have contributed to these outcomes.

THEORETICAL FRAMEWORK

Our theoretical framework focused on two core ideas that functioned in the Urban Math project as guiding perspectives as well as subjects of research:

symbolic places in mathematics learning and family resemblances across learning experiences in mathematics.

Symbolic Places in Mathematics Learning

Building on the profound similarity between the ordinary processes of becoming familiar with a new place (e.g., a town, a train station) (Greeno, 1991) and the process of learning a new mathematical concept, we proposed that learning mathematics entails becoming familiar with "symbolic places." Unlike physical places, symbolic places can involve entities that do not exist in a physical way, such as mathematical functions or geometric shapes. Symbolic places can, however, be made visible through symbols or inscriptions (e.g., graphs, equations, diagrams), similar to the ways an invisible city can be portrayed by a map. Often physical places completely preexist our encounter with them, but sometimes we create them as we encounter them (e.g., moving to a new apartment and bringing furniture, appliances). Symbolic places probably have more of the latter quality (i.e., we make them as we dwell in them) although frequently we experience them as fully preexisting in a Platonic fashion. Notwithstanding the differences between physical and symbolic places, we claim that the ways in which we get to know them are deeply similar. Many of the features that help us to get a rich sense of a physical place are also important in "inhabiting" symbolic places: being able to wander, talk with others about how they see the place, contrast personal images with known stereotypes, share our surprises, be advised on what to expect and how to be prepared, use guides and maps, and so forth.

Crucial to our notion of symbolic place is the blurring of rigid boundaries between the learners and the learning environment. Such blurring is analogous to a current trend in biology, which questions the traditional dichotomy between organism and environment, according to which the latter is a "given" to which the organism adapts (Lewontin, 2000; Turner, 2000). Lewontin (2000), in reviewing how organisms create local environments, which can be radically different from the surroundings apparent to an external observer, was led to postulate, "If one wants to know what the environment of an organism is, one must ask the organism" (p. 54). Similarly, we think that teachers and students constitute and inhabit symbolic places that can be quite different from what observers might expect from examining the sequence of classroom activities, and that grasping the symbolic places in which they dwell entails putting aside our pre-conceptions and listening receptively to what they tell us. Similar to the ways biologists are being prompted by the microanalysis of life to question what adapts to what (i.e., the organism to the environment or vice versa), we are recognizing that a learning environment is codeveloped by the student, the tasks, the tools, the classroom conversation, and so forth, which powerfully supersedes the notion of a learning environment as one given to the students and to which each student strives to adapt.

Family Resemblances Across Learning Experiences in Mathematics

The old notion that understanding is an all-or-none phenomenon is linked to the idea that each concept has an essence, customarily expressed in its definition, so that its understanding is ultimately a matter of possessing or not possessing such a definitional essence. Advancing the alternative perspective that understanding is *not* an all-or-none phenomenon entails articulating a competing view of what concepts are. In order to do so, we have drawn on Wittgenstein's (1953) idea of "family resemblances" and Rosch's (1999) views of concepts as participatory. We postulate that mathematical and scientific concepts are developed by participating in or inhabiting many symbolic places and extending family resemblances across them, as noted in Noble, Nemirovsky, Wright, and Tierney (2001):

> We reflect on how one understands the familiar idea of one half. Try out these two tasks yourself. First, find one half of the quantity 3275 and write your answer. Next, walk across a room once, and then walk across the room again, but try to walk half as fast as you did the first time. These two examples clearly share the mathematical idea of dividing something in half, but probably your experiences of dividing something in half had many differences as well as similarities. Dividing 3275 in half most likely drew on your computational ability and your number sense, causing you to think about one half as a relation between two numbers. Walking across the room half as fast as before probably caused you to think about how to qualitatively compare one or more of the quantities speed, distance, and time when comparing your two motions across the room. You may also have tried to feel the quantities of speed or time in your body as you made the motion itself. In these ways you may have related two experiences of speed to arrive at one half, or you may have perceived the continuous quantities of time or speed and compared them.

> Although one may be tempted to look for the essential element of "halfness" in each of these experiences and to try to find ways to give students access to this element, one's own sense of halfness comes from these experiences and many others like them. The similarities and the differences among these experiences of *one half*, and all the experiences of *one-half* relationships one has had in a lifetime constitute the family of *one half* that allows one to recognize what are *one-half* relationships and what are not. (pp. 105–106)

This analysis allowed us to draw pedagogical implications for our perspective versus relying on conveying the essence of a given idea to students, as in the essentialist perspective. We have illustrated these implications in Noble et al. (2001) by using Wittgenstein's (1953) famous example of the meaning of the word *game*:

> A teacher with an essentialist perspective on the meaning of *game* would give great prominence to the definition, because the definition would determine general criteria for which activities belong to the class of "games" and which do not. The pedagogical approach might be to show examples of games and not-games and to indicate the presence or absence of the defining features in each example. Or one may first give examples, and let students discover the definition. In both cases, the definition is the target of the lesson: Once the student correctly applies the definition to categorize activities, one would say that he or she has understood what a game is. On the other hand, if one takes the perspective that developing family resemblances across lived-in spaces is critical to learning, then the most important aspect of learning and knowing what a game is becomes one's life experience with games. On the basis of the kinds of games one has participated in or watched and one's impressions of these games, one recognizes certain activities as being of the family of games and others as not of this family. On that same basis, one assesses whether some definitions of *game* are better than others. From this point of view, the pedagogical focus would be on allowing and encouraging students to participate in a range of games, to make these games into lived-in spaces for themselves, and to reflect on their experiences of playing games, through discussing, symbolizing, and comparing games.

METHODOLOGY

Levine (1983) introduced the phrase "the explanatory gap" to describe an open range of phenomena that cognitive science, at least in 1983, did not account for: how the world is experienced by human beings. Cognitive scientists have postulated many types of mechanisms for, say, visual perception, but these mechanisms are largely silent when it comes to what it is to be conscious of an object in front of us or to attribute meanings to our surrounding space. The literature on the explanatory gap often cites Nagel's (1970) paper, "What Is It Like to Be a Bat?" The question "What is it like?" is a key to the explanatory gap: What is it like to be a student dealing with this or that problem? What is it like to participate in these classroom activities? How do things look and feel to a certain student? Bridging the explanatory gap entails a richer understanding of phenomenality, of the ways things appear to someone in a particular situation. Understanding phenomenality is particularly central in education because good educators are constantly aware of the need to figure out how things look to their students.

Many scholars have developed empirical approaches to investigating phenomenality, particularly in the area of anthropology and cultural studies (Geertz, 1973) but also in educational research (Lincoln & Guba, 1985). In the area of medicine, some studies, like those by Oliver Sacks (1995, 1998) on patients with neurological diseases attained a certain notoriety because they offer us a glimpse into strange and extraordinary views of the world. Because our goal was to contribute to an understanding of the phenomenality of the bodily and the symbolic in the context of mathematics learning, we were concerned with developing methodology appropriate to studying phenomenality in mathematics education (Nemirovsky, 1994; Nemirovsky & Monk, 2000; Nemirovsky, Tierney, & Wright, 1998; Noble et al., 2001). Our approach to the analysis of the videotaped classroom episodes shares a number of commonalities with Interaction Analysis, as described by Jordan & Henderson (1995), and the interpretive approach described by Packer and Mergendoller (1989). Rather than approaching the videotaped classroom episodes with a predetermined coding scheme, we allowed the analysis to "emerge from our deepening understanding" of the events unfolding on the videotaped record (Jordan & Henderson, 1995). We treated the participants' utterances and actions as the results of processes accomplished over time and in interaction with others, and we focused our attention on the details and meanings of these actions and utterances (Packer & Mergendoller, 1989). Our data analysis took place in a group of four researchers with varying interests and expertise who continually challenged each other's observations and required of each other the grounding of interpretations in observable events on the video record (Jordan & Henderson, 1995).

WHAT WE HAVE LEARNED

In order to describe what we have learned through our studies, we have selected three classroom episodes (from three different high school math classes) that provide a basis from which to elaborate on the ways that classrooms *focused on student understanding* develop mathematical symbolic places and family resemblances. These episodes are focused on three themes: (a) the emergence of a sense of direction in a classroom mathematical conversation, (b) the continuity between everyday and technical language, and (c) the role of the interplay between physical phenomena and symbolic expressions.

THE EMERGING SENSE OF DIRECTION

There is a widespread tendency to discuss and describe teaching as a set of choices among polarized options: "telling" versus "not telling," teacher-centered versus student-centered, skills versus concepts, and so forth. In agreement with Chazan and Ball (1999), who argued that the opposition between telling and not telling is a false one, we have come to think that teaching involves allowing a sense of direction to emerge from the classroom interaction, which is constructed jointly by the students and the teacher in ways that cannot be aligned along such dichotomies. Often teachers depart from the execution of preconceived plans in order to be responsive to questions, concerns, and possibilities that could not have been predicted. Furthermore, many of the so-called *teacher moves* are expressions of unspoken feelings, which only in retrospect can be made to fit into a neat sequence of *interventions*. This is consistent with Simon (1995), who suggested that, although teachers plan lessons and design tasks aimed at realizing hypothetical student learning trajectories, in actual interaction with students, teachers constantly revise and redefine their plans. This departure from seeing teaching as execution of plans or choices between dichotomies does not mean that nobody makes decisions or that the participants are unaware of what they are doing or saying. Instead, it means that plans, agendas, goals, and so on become part of a complex dynamics from within which ideas and threads emerge as a confluence of multiple perspectives and sudden insights. We call such merging of perspectives and insights a "sense of direction." Perhaps a good metaphor for this instructional activity is that of hiking across an unknown forest, given certain constraints (e.g., time), tools (e.g., a trail map), and destination while being open to changing course if it becomes advantageous or appealing. Although the teacher is decisive in fostering a sense of direction, students often assume a leading role or contribute ideas that change the orientation of the whole conversation. We see teachers as working on a sense of direction for themselves as well as for their classes. At times, this work is expressed through efforts to preserve certain voices that otherwise might vanish, to foster mutual listening, and to comment on the fruitfulness of the different contributions.

The Episode

This episode took place in a Jesse Solomon's classroom of 10th and 11th graders in a second-year integrated mathematics course at City on a Hill public charter high school in Boston, MA. City on a Hill admits students by lottery from across the city of Boston; approximately 80% of the students are students of color, and over half the students qualify for free or reduced lunch. (For a more extended analysis of this episode, see Solomon & Nemirovsky, in press).

At the start of the class, Mr. Solomon asked the students whether they had had difficulties with the homework, intending to devote a few minutes to reviewing problems. Discussing any particular problem for an extended time was not part of his lesson plan for that day. The selected 16-minute discussion followed a question posed by one student (Maria) about a problem that she hadn't been able to solve. Problem 18 (Rubenstein, Craine, & Butts, 1995), gave a sequence of four numbers (1, 8, 27, 64, ...); students were to graph at least six points and decide whether the sequence appeared to have a limit.

Maria started the discussion by saying:

> I couldn't figure out what the next two [numbers] would be. I figured—I did the differences, I did multiplication, I tried times 2 plus 4, times 2 plus 3. I tried that. I couldn't get it. My mom couldn't get it, and she told me, just clean my room.

Students started to brainstorm possible rules including power and linear relationships (e.g., $8 \times 8 = 64$; $(1 + 8) \times 3 = 27$). Mr. Solomon recorded their ideas on the overhead projector. This interaction had complex emotional and interpersonal nuances. Although all ideas were accepted for consideration, some were refuted with irony, and others engendered laughter led by the authors themselves. At one point, Mr. Solomon asked the group whether they wanted him to tell them the fifth term in the sequence. The group was divided: Some students felt that they should be able to get it by themselves. One student, Jamal, argued that he had the solution, but he would not tell it in order to "let others think." After being pressured by the students and Mr. Solomon, he agreed to tell it. He started, "The difference between the differences between ...": Jamal thought that the solution was based on successive differences (see Fig. 8.1a).

Jamal soon realized, however, that he had made a mistake (he had miscalculated one of the first differences as 21) and that his idea, therefore, would "not work." With the first differences now on the transparency, however, other students started to play with patterns, such as the second digit of each difference being 7, 9, 7, ... and the first digit being the sequence of odd numbers, or the previous one + 1, + 2, + 3, and so on. This is a typical occurrence in mathematical conversation: Mistaken ideas often inspire new approaches.

Several students gave their opinions along the way on the value of the ideas offered and of the conversation itself (e.g., whether Mr. Solomon should tell the fifth number, requests to "move on" when certain approaches seemed unfruitful, assertions that four numbers were too few to work with). Students proposed different fifth numbers for the sequence: 121, 133, 113, and 123. Naomi made the case for 123 by postulating that the 4th first difference would be 59 because the second digit alternated between 7 and 9 and the first digit was the sequence of odd numbers (so that $59 + 64 = 123$). This defense of 123 was a turning point of the conversation

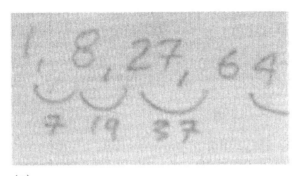

(a)

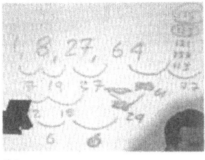

(b)

FIG. 8.1 Student strategies to find the 5th term of a series: "The differences between the differences between ..." (a) Looking at the 1st difference between terms in the series. (b) Comparing 2nd and 3rd differences in the series.

because hitherto the class had held a tacit assumption that there was a unique "right" number, known by Mr. Solomon. This assumption was also held by Mr. Solomon, in the sense that he had 125 (5^3) in mind as the target value and would have told the class that number if they had asked. Mr. Solomon responded to Naomi's argument for 123 by saying, "OK, that could work." His assertion marked an important shift for the class into acknowledging the multiplicity of possible sequences, given those four numbers.

Next, Nadia wanted to make the case for her proposed answer of 121. She started by asserting: "I didn't stop at finding the difference between the numbers given. I found the difference between the differences." Apparently Nadia had generated 121 by assuming that the third difference (6) remained constant. After correcting an arithmetic mistake pointed out by Mr. Solomon, she changed the 121 into a 125 (see Fig. 8.1b).

The bewildered reaction from some students was immediate: "This is ridiculous! Differences from differences!" "This is crazy." Mr. Solomon recognized the connection between polynomial powers and the constancy of successive differences, but noted in his journal:

Although I knew something about the use of differences in determining the kind of polynomial equation, these ideas were not fresh in my mind, and I was not completely confident that I knew how or why a constant third difference indicates a cubic function. It may have taken until the end of this episode for me to see clearly.

He then explained to the class that even though he (or the textbook) had an answer in mind, there were infinitely many possible rules to continue the sequence, then noted that the set 1, 8, 27, 64 is a familiar one, suggesting to many people the rule $1^3 = 1$, $2^3 = 8$, $3^3 = 27$,.... Several students reacted with "of course" ("Oh, yeah, I knew it!"); then Mr. Solomon pointed out the mathematical significance he had recognized in what was written on the transparency:

Mr. Solomon: If you took two sets of differences, and found that all these numbers are coming up the same, you would then know that it's a quadratic function, that it's a second degree function.

Jamal: Ooohhhh! [pause] So if this fourth one, if you had to take a fourth one, it would be, like, a 4th power?

Commentary

Neither the textbook sequence for the lesson nor Mr. Solomon's plan for this day involved elaborating on and engaging the students with successive differences of power sequences. This outcome took shape as a manifestation of three different aspects of this classroom culture: the teacher's openness to student contributions, a classroom culture that valued brainstorming, and the teacher's recognition of the emergence of something mathematically important.

Mr. Solomon's Openness to His Students' Contributions. Although the conversation started with a shared assumption that Mr. Solomon knew "the" solution to the problem, the exchanges never fell into the game "guess my mind," which is often the main form of interaction between teachers and students. Mr. Solomon and the students treated his knowing the expected fifth term of the sequence as a resource, the use of which was subject to students' control. By avoiding the "guess my mind" dynamics, Mr. Solomon engaged his students in a genuine conversation, thereby opening students' exploration. This was most notably visible in his comment "OK, that could work"—admitting to himself and to the students that they had been talking as if the textbook answer (and Mr. Solomon's) was the single correct way of continuing the sequence.

Being open to the students' contributions does not mean that the teacher is "neutral" and forced to hold back his or her own valuation of the ideas being offered. In point of fact, this would be impossible. Different ideas are generated simultaneously, and preserving or highlighting certain voices and approaches necessarily shifts other voices and approaches to the margins, at least temporarily. In addition, creating a collective space to articulate a particular idea influences the course of other streams of thought in ways that cannot be fully anticipated by the participants.

Finally, being open to students' contributions is not incompatible with direct explanations and explicit telling. The interest with which students listened, as well as their comments on Mr. Solomon's explanation, reflected how significant that connection between powers and constancy of the differences had become for the class. This connection pulled together many of their questions and contributions and enabled Mr. Solomon to validate their thinking as having touched on important mathematical matters.

A Classroom Culture That Valued Brainstorming as a Way to Solve Problems, Students' Attempts at Figuring Problems out by Themselves, and the Fruitfulness of Following Different Paths. Many ideas offered by the students ended up being dismissed or inconsequential. Crucial to the conversation was the attitude, embedded in this classroom culture, that being mistaken or putting aside a certain course of action was not "damaging" to the contributor, unless he or she became an obstacle to the group's moving forward (something felt at one point with respect to Jamal's refusal to tell what he "knew" to be the answer). Even though the conversation included many expressions of irony and criticism (e.g., "Challenge yourself!"), proposals were not considered or set aside solely on the basis of a student's being considered good at math or the force of his or her personality.

Several students made comments about the flow of conversation (e.g., "Now we are forcing it") and the potential (e.g., "I know why you said 121") or strangeness (e.g., "This is crazy") of certain contributions. Such commentaries about the conversation itself suggest that the teacher was not the only one trying to develop a sense of direction. To the contrary, some of the students become active co-developers of the sense of direction that emerged.

Mr. Solomon's Recognition That Nadia Was Hitting on Something Mathematically Important. So-called teacher moves are often expressions of unspoken feelings. Mr. Solomon worked to correct several arithmetic mistakes to preserve the development of the "difference of the differences" idea, even though he was not fully clear at the time on "how or why a constant third difference indicates a cubic function." We can see this as an example of the role of the teacher's mathematical competence. Being fluent and knowledgeable about mathematics proved critical to recognizing, on the spot, a rich sense of direction, in spite of not having the content "fresh in my mind" or answers prepared in advance for students' spontaneous questions.

Epilogue

In this example, we portrayed the complexity involved in managing the wide range of ideas and choices emerging from the interactions among students and the teacher. In describing such "management" in this case, we postulate that the teacher, in conjunction with some of the students, developed an emergent *sense of direction* responsive to what they intuited as valuable and mathematically significant in light of the proposed contributions. This example illustrates the importance (to the emergence of a rich sense of direction) of the teacher's fluency with a wide range of mathematical ideas, a classroom culture in which mistaken ideas can be inspiring, and the teacher's fostering of genuine openness throughout classroom interactions.

THE CONTINUITY BETWEEN EVERYDAY
AND TECHNICAL LANGUAGE

Educators are often highly concerned about the precise use of mathematical terms and the role of definitions of words in mathematical lessons. Many have raised related concerns about the "danger" of using certain words because of their dissimilar meaning or imprecise use in everyday language. Pimm (1987) proposed that knowing mathematics involved fluency with a "mathematical register": a sort of a dialect or linguistic subgenre in which words have meanings different from the ones conveyed in common use. Because of these concerns, many researchers have attempted to understand the relationship between everyday language and the mathematical register. Moschkovich (1996), for example, described complex negotiations of meanings between pairs of students describing graphical lines corresponding to different linear equations. In these conversations, the students tried to ascertain how to use common words such as *steep* or *slope* within their mathematically accepted meanings. Extending this body of work, we found that the relationship between everyday language and the mathematical register is one of continuity. This continuity implies that the mathematical meanings are normally expressed in the combined use of everyday language and the specialized register and that, far from being "dangerous," the open discussion of multiple word usages, including those that tend to be mistaken from a mathematical point of view, enriches mathematics learning.

The Episode

This episode took place during the first unit of an AP calculus class taught by Marty Schnepp at Holt High School located just outside East Lansing,

MI. The curriculum for this class, developed by Mr. Schnepp over a period of 8 years, owes much to his close attention to his students' ideas and his own and his students' intuitions. The course begins with a series of activities involving numerical integration. For 3 weeks prior to the episode, students had been developing numerical integration. In the few classes immediately previous to this episode, they had been using a pair of "minicars," one red and one blue moving along two linear and parallel tracks, the velocities of which were defined by algebraic equations typed into a computer. This tool is an example of LBM (Line Becomes Motion) technology, which allows students either to control physical phenomena by defining mathematical functions on the computer or to generate functional graphs by moving physical devices by hand. (More detailed analysis of this episode can be found in Schnepp & Nemirovsky, 2001.)

During the previous session, the students had been discussing ways to calculate the average speed of the minicars; the discussion ended without resolution. Attempting to revisit the topic of average speed, Mr. Schnepp started the lesson by asking:

> If the red car traveled according to $v(t) = .1(t - 11)^2 - 2$, and the blue car moved at a constant rate, how fast would the blue one have to go in order to start and stop in the same place?

His idea was to have the students figuring out the constant speed of the blue car so that both cars would start and finish simultaneously next to each other; in such an arrangement, the constant speed of the blue car would be the average value of the function driving the red car over a well-defined time interval. Students worked in groups on this problem for about 10 minutes, and then Mr. Schnepp called them together. From several suggestions, the class chose a value of 1.5 cm/s for the constant speed of the blue car. The minicars were then set so that the blue car would move at that speed, and the red car would move according to the function given. The result was that the cars appeared to travel the same distance (see Fig. 8.2).

Mr. Schnepp asked, "So is this the idea that you guys were talking about when you were talking about average velocity [the previous day]?" After several affirmative responses, he asked, "So what's a good definition [of *average velocity*]?" One student, Jan, apparently reading from her paper, suggested a definition of *average velocity*, which Kim wrote on the board. After input from other students (rate replaced speed, object replaced minicar, and displacement replaced distance), the final form of Jan's average velocity definition was:

> The rate that the object would have had to had constant to achieve the same displacement in the same amount of time.

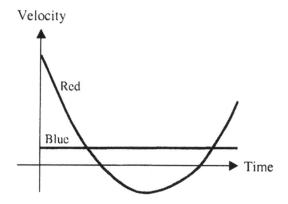

Red car: $v(t) = [.1(t - 11)^2 - 2]\ cm/s$

Blue car: $v(t) = 1.5\ cm/s$

FIG. 8.2   Calculating the average velocity of a car (moving at a velocity determined by a velocity formula) over a displacement determined by its "intersection" with a car moving at a constant velocity.

This prompted the following exchange between students:

Sam: I'm having trouble with this negative velocity stuff. See, 'cause, seems like for this definition, if we were to, like, [pause] if the car were to go, like, 5 cm/s forward and 5 cm/s back, for the same amount of time and got back to where it started. We would be saying that its average velocity was zero....

[He goes on to say that negative velocity only means direction. He concludes this is not relevant to the average and that, for his example, the average velocity should be 5 cm/s.]

Sue: So, are you basically trying to say, is that you'd want the average velocity to equal out to the same distance in the end, same distance traveled? [Pauses] Because then you'd ignore the direction of the velocity and make everything positive. [Pauses] I did that.

Sam: That would make more sense to me, I think what we have here [in the definition] is average speed ... for the displacement.

Another student had worked with Sam. He was asked to come up and draw a graph that Mr. Schnepp had seen on his paper relating to the issue of "making everything positive." The student drew the graph shown in Fig. 8.3, saying, "I just did absolute value" and "it would cover the same distance.... It wouldn't end at the same spot in the same time. It would cover the same amount of distance." At that point, all students agreed that a rate between 2.4 and 2.6, if maintained for 20 seconds, would cause the blue car to travel the same total distance as the red car, as obtained by the absolute value of the velocity function (i.e., the "reflected" function shown in Fig. 8.3) but not end in the same location.

Students further agreed on the significance of two distinct concepts: a constant rate that would result (a) in the same displacement (or change in position) or (b) in the same total distance. A discussion on the distinction between speed and velocity ensued. They argued the question, "Which [quantity] can't be negative?" Eventually, Rob suggested a distinction of "distance over time" for speed and "displacement over time" for velocity. Sam argued the opposite. Rob continued, saying that the $|v(t)|$ graph (see Fig. 8.3) would result in an increasing change in distance for the "reflected region," whereas the other graph (see Fig. 8.2) would "accumulate a decrease" in the region where $v(t)$ had negative values. This was because the moved backward during that time period. Kim pointed out that, because a car's speedometer cannot have negative values, she believed that speed is always positive. This seemed convincing to everyone, but the debate, though quieted, continued.

Students left class with an assignment to sketch an accumulation graph for the $v(t)$ graph. The following day, all four graphs—speed, total distance, velocity, and displacement—were discussed. All these subtleties became part of their distinctions between average speed and average velocity.

Commentary

Sam's and Sue's reactions to the definition of average velocity (written on the whiteboard by Jan) expressed their perception that such definition misses something important. Saying that the average velocity of a car that

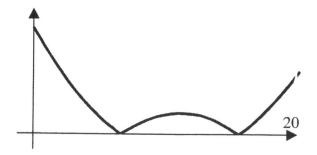

FIG. 8.3 Using the absolute value of the velocity of a car (moving at a velocity determined by a velocity formula) to determine distance traveled, regardless of direction.

gets back to where it started is zero might be a correct assertion, but it is exceedingly "uninformative": It gives no sense of how far or how fast the car actually moved. In contrast, averaging speed using the distance traveled but ignoring direction seems to provide a richer idea of the motion of the minicar over that period of time. Both computations are viable measures of motion, but each offers a different slice of information about that motion. Note that the whole distinction was rooted in what the class alluded to as the difference between "same displacement" and "same distance traveled." Students' understanding of what these two phrases refer to and their fluency in using them is part of their fluency with everyday language. In customary English usage (as well as in its etymological origins), the word *displacement* has the connotation of a disparity between two locations. The emphasis of "displacement" is on difference of place whereas the emphasis of "distance traveled" is on the length of travel between these locations (represented schematically in Fig. 8.4).

Grasping technical terms, such as average *velocity* and average *speed*, is not a matter of putting aside the usage of everyday language and starting to use words solely according to their formal definitions. Everyday language can be picked up, extended, ignored, or distorted by technical usage (and vice versa) and is a vast and inescapable source of word meanings. When Sam said, "I think what we have here [in the definition] is average speed ... for the displacement," he was indicating this interpretation as one possibility among others that everyday language affords us. Locating a technical meaning in consonance and dissonance with other possibilities arising from the richness of everyday language marks an indispensable continuity, and one full of tensions and ambiguities. This episode typifies how from the midst of these tensions and ambiguities, and not from self-contained definitions, the "mathematical register" becomes a live dialect rather than a dead jargon destined to be forgotten.

Epilogue

Our studies have documented that as students acquire fluency with technical terms, they do not stop talking everyday language, but instead learn to

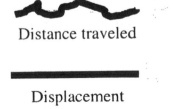

Distance traveled

Displacement

FIG. 8.4   Distinguishing displacement (distance from starting place to ending place) from distance traveled (length of path taken between starting and ending places).

make distinctions among the multiple meanings latent in both everyday and technical language. We see classroom conversations that support the learning of mathematics as those in which students are encouraged to describe mathematical patterns in many ways, including using words they are familiar with. Rather than insisting that students' talk be confined to prescribed technical terms and definitions, we believe that formal definitions become meaningful and productive when students are able to redescribe them in common language, to offer multiple examples, and to employ various ways of talking about them. This premise is as valid in bilingual as in English-speaking classes. Given that each language has idiosyncrasies and specific types of word uses, the hypothesis of continuity suggests that students who are not native English speakers can produce unique contributions to this desirable diversity in classroom talk, as shown in the next example.

PHYSICAL TOOLS AND SYMBOLIC EXPRESSIONS

From Cuisenaire rods to conic sections, from Dienes blocks to the Lenard sphere, manipulatives are and have been important in mathematics education, and the current availability of computer-simulated versions has not eliminated physical manipulatives from commercial catalogs. Manipulatives are also popular with many mathematics teachers, who feel that they can play significant roles in learning mathematics. But this intuited helpfulness is not well understood. Many math educators (Ball, 1992; Lesh, Post, & Behr, 1987) have elaborated important criticisms of the idea that doing math problems with concrete materials—the "hands-on" hypothesis—automatically determines a higher level of understanding. These critics argued, correctly we believe (Noble et al., 2001), against the assumption that the mathematics to be learned is "in" the materials, with the consequent assumption that by manipulating these materials the students will come to see the mathematics therein. Many have instead argued that students constitute the materials through their activities and talk (Meira, 1998), in ways that might (or might not) have a significant connection to the mathematics they are expected to learn.

The hands-on hypothesis tends to lose sight of the anticipatory nature of understanding. Perception and cognition are inherently predictive, allowing us to anticipate the consequences of current or potential actions (Berthoz & Gielen, 1993). We move our eyes on the basis of what we anticipate seeing—we do not look at anything without some anticipation of what it could be. Anticipations are constantly revised and adjusted on the basis of what we do see. By this, we do not mean anticipating in the sense of thinking, before looking, about what it is that we are going to see, but in the sense of looking somewhere to address questions that arise in us, questions that often come to us unexpectedly, prompting quick gaze shifts. This suggests that what students see in manipulatives is related to what they anticipate. Without entertaining mathematical conjectures and letting mathematical questions emerge, there is no mathematics to be found in the manipulatives. This implication, then, is that there is no deterministic outcome associated with using manipulatives; however, when they *are* part of anticipatory conjectures and reflective intentions, their use can bring in aspects of the human experience, such as tactile exploration, kinesthetic engagement, and multiplicity of points of view, which might otherwise be absent.

The Episode

This episode took place in Apolinario Barros' Algebra I class at the Jeremiah E. Burke High School in Boston. Mr. Barros teaches bilingual classes in Capeverdean Creole; most of his students immigrated to the United States from Cape Verde during the 2 years previous to the episode.

On the first day of a series of activities on quadratic functions and parabolas, the class was using minicars similar to the ones employed in the previous episode. Before the students entered the classroom, a parabola was designed in the software, which was later used to run the red car, and the screen was covered. When class started, students were asked to observe the motion of the car governed by the hidden parabola (see Fig. 8.5). The students then drew rough graphs of the motion they had observed and posted their drawings at the front of the class.

One of the students, Cara, started describing the motion of the car, as she remembered it, to the class. She mentioned that the car had started to slow down before it reached the return point. The following exchange took place:

Edmond: [To Cara] How did you manage to see that?
Alan: With her eyes!
Cara: With my eyes! [Laughter]

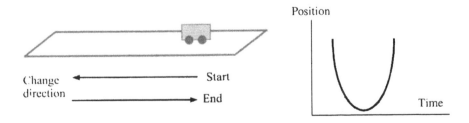

FIG. 8.5   Minicar moved by a parabola generated on a computer (screen image).

Edmond: With your eyes you don't know. [Pauses] Sometimes, some-
             times the car slows down, and you think it has the same speed.
Cara: Edmond, if you looked carefully over here, it clearly
             changed speed. You can see that right away.

The class discussed differences between the posted graphs. (Some of
these are shown in Fig. 8.6.) Alfredo's graph had a little horizontal segment
on the low vertex (like the graph on the left) whereas Cara and Beatriz's
graph had a curved vertex (like the graph on the right).

Then the class discussed how the car's slowing down was shown by the
curviness of the graph. Alfredo raised a new point:

Alfredo: You [Cara] did not say that speed ... We saw that when the car
             went back, it stopped a little. We must represent on the graph
             how long, how much the car stops. The car was going like this
             [see Fig. 8.7a]. As Cara said, it decreases speed [moves his
             hand forward]. It stops a bit [stops]. Then it continued [moves
             his hand back]. Now, you must represent on the graph [points
             to the graphs] how much the car stopped. I showed it over
             there ... I put a line like this [horizontal dotted line; see Fig.
             8.7b] where it stopped and then another [dotted line going
             up]. So the straight line represents how long the car stopped
             ... If you show the graph to someone ... If you show it [the
             graph] to someone, Barros, without anyone saying anything,
             they would not know that the car stopped.
Mr. Barros: But the car stopped?

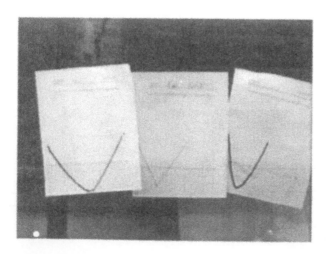

FIG. 8.6 Students' rough graphs of the function describing the motion of the minicar.

Alfredo: Nobody would know that the car stopped. You have to show that the car stopped over there.

Mr. Barros: Cara has an argument.

Cara: Right here [where the car turned back], it did not stop. It just touched and went back. It doesn't stop....

Beatriz: She [Cara] doesn't want to say that the car stops [makes a stopping gesture with both her palms facing the table] ... Yeah, we don't want to say that the car stops [hits fist on the table] and stays there and then goes. We want to say that it changes speed when it goes backwards [moves her hands towards herself, simulating the motion of the car] and then when it goes forwards [points to the tracks and makes a forward motion with her hand].

Alfredo: Barros, Barros, can you put the cars one more time please?

Mr. Barros: I'll put them on in a bit, but I want to hear more comments.

(a)

(b)

FIG. 8.7   Alfredo's gestures illustrating minicar motion. (a) Hand moving forward and back to suggest apparent stop before reversal of direction. (b) Hand movements showing the flat-bottomed parabolic shape he drew to show car motion.

Edmond:	Mr. Barros, Alfredo's graph shows that the car stops so many minutes—
Cara:	Yes—
Edmond:	How many seconds, 1, 2, 3 seconds. But the car doesn't stop. The car comes, then it turns. For example, if a car is going around a garden, it turns [makes several circular motions with hand] and takes off.
Cara:	Yeah. Barros?
Beatriz:	Yeah.

Commentary

Right after Cara described the car's motion as slowing down on a certain stretch, Edmond questioned her on her certainty. Edmond was not arguing whether such slowing down had taken place, just that by simply looking it was hard to know. Cara, however, felt that by looking carefully "You can see that right away." Alfredo, on the other hand, questioned Cara and Beatriz on a different issue: He had seen the car stopping for a short time before moving back to the start position and thought that this stop had to be reflected with a little horizontal line at the vertex, which was absent in Cara and Beatriz's graph. He added that anybody looking at their graph would miss that component of the car's motion. Cara and Beatriz responded by claiming that the car did not really stop: "It just touched and went back." This latter argument was also brought up by Edmond, who then introduced the example of a car going around and turning back without ever stopping.

The students in this episode were discussing physical phenomena (e.g., Did it slow down? Stop?) and their corresponding symbolic expressions (e.g., How to show the car turning around? Would someone else understand?). The physical cars driven by graphs of functions prompted students to shift from the usual "This is how things are" to "This is how I [or you or she] see things." In other words, the shared device served as a common object of study, helping students to sharpen and articulate their differences in perspective. The process of telling each other their views of the motion they had all seen engaged the students with the details of graph interpretation (e.g., the meaning of lines vs. curves, of flat vs. slanted lines).

This example captured what has become for us a key role of manipulatives and physical devices in mathematics learning: to support the emergence and articulation of different ways of understanding phenomena that are available for common observation. This is different from the hands-on hypothesis that assumes that the device somehow conveys a piece of mathematics. Instead, the mathematics in this episode, centered on the interpretation of graphs of quadratic functions, was constituted by the anticipatory

acts of seeing and symbolizing—anticipatory aspects that get expressed in almost all the utterances (e.g., "If you looked carefully over here, it clearly changed speed"; "If you show it to someone, Barros, without anyone saying anything, they would not know that the car stopped").

Epilogue

Manipulatives are used almost exclusively in the lower grades and are seen as being suitable for young children who "have concrete thinking." In addition, often they are assumed to "show" the mathematical idea, as if there were a deterministic relationship between handling these manipulatives and seeing the concepts that mathematically knowledgeable people think they embody. Our studies support the notion that there is a place for physical materials at all levels of mathematics education and that they are helpful not in themselves, but as part of rich mathematical conversations, through which students can compare their different ways of perceiving shared objects and events.

CONCLUSION

Through teaching experiments in three high school mathematics classes, the Urban Math project identified important processes that take place in classrooms for understanding: the ongoing emergence of a sense of direction, the continuity between everyday and technical language, and the role of physical materials as "pieces of conversation" about shared objects and events.

REFERENCES

Ball, D. L. (1992). Magical hopes: Manipulatives and the reform of math education. *American Educator, 16*(2), 14–18, 46–47.
Berthoz, A., & Gielen, C. (1993). *Multisensory control of movement.* Oxford, UK: Oxford University Press.
Chazan, D., & Ball, D. L. (1999). Beyond being told not to tell. *For the Learning of Mathematics, 19*(2), 2–10.
Geertz, C. (1973). *The interpretation of cultures.* New York: Basic Books.
Greeno, J. G. (1991). Number sense as situated knowing in a conceptual domain. *Journal for Research in Mathematics Education, 22*(3), 170–218.
Jordan, B., & Henderson, A. (1995). Interaction analysis: Foundations and practice. *Journal of the Learning Sciences, 4*(1), 39–103.
Lesh, R., Post, T., & Behr, M. (1987). Dienes revisited: Multiple embodiments in computer environments. In I. Wirszup & R. Streit (Eds.), *Developments in school mathematics education around the world.* Reston, VA: National Council of Teachers of Mathematics.

Levine, J. (1983). Materialism and qualia: The explanatory gap. *Pacific Philosophical Quarterly, 64,* 354–361.

Lewontin, R. (2000). *The triple helix.* Cambridge, MA: Harvard University Press.

Lincoln, Y. S., & Guba, E. G. (1985). *Naturalistic inquiry.* Beverly Hills, CA: Sage.

Meira, L. (1998). Making sense of instructional devices: The emergence of transparency in mathematical activity. *Journal for Research in Mathematics Education, 29*(2), 121–142.

Moschkovich, J. N. (1996). Moving up and getting steeper: Negotiating shared descriptions of linear graphs. *Journal of the Learning Sciences, 5*(3), 239–277.

Nagel, T. (1970). What is it like to be a bat? *Philosophical Review, 79,* 394–403.

Nemirovsky, R. (1994). On ways of symbolizing: The case of Laura and the velocity-sign. *The Journal of Mathematical Behavior, 13,* 389–422.

Nemirovsky, R., & Monk, S. (2000). "If you look at it the other way ...": An exploration into the nature of symbolizing. In P. Cobb, E. Yackel, & K. McClain (Eds.), *Symbolizing and communicating in mathematics classrooms: Perspectives on discourse, tools, and instructional design* (pp. 177–221). Mahwah, NJ: Lawrence Erlbaum Associates.

Nemirovsky, R., Tierney, C., & Wright, T. (1998). Body motion and graphing. *Cognition and Instruction, 16*(2), 119–172.

Noble, T., Nemirovsky, R., Wright, T., & Tierney, C. (2001). Experiencing change: The mathematics of change in multiple environments. *Journal for Research in Mathematics Education, 32*(1), 85–108.

Packer, M. J., & Mergendoller, J. R. (1989). The development of practical social understanding in elementary school-age children. In L. T. Winegar (Ed.), *Social interaction and the development of children's understanding.* Norwood, NJ: Ablex.

Pimm, D. (1987). *Speaking mathematically: Communication in mathematics classrooms.* London: Routledge and Kegan Paul.

Rosch, E. (1999). Reclaiming concepts. In R. Nunez & W. J. Freeman (Eds.), *Reclaiming cognition* (pp. 61–77). London: Imprint Academic.

Rubenstein, R. N., Craine, T. V., & Butts, T. R. (1995). *Integrated mathematics* (Book 3). Boston: Houghton Mifflin.

Sacks, O. W. (1995). *An anthropologist on Mars: Seven paradoxical tales.* New York: Knopf.

Sacks, O. (1998). *The man who mistook his wife for a hat and other clinical tales.* New York: Simon & Schuster.

Schnepp, M., & Nemirovsky, R. (2001). Constructing a foundation for the fundamental theorem of calculus. In A. Cuoco & F. R. Curcio (Eds.), *2001 yearbook: The roles of representation in school mathematics* (pp. 90–102). Reston, VA: National Council of Teachers of Mathematics.

Simon, M. (1995). Reconstructing mathematics pedagogy from a constructivist perspective. *Journal for Research in Mathematics Education, 26*(2), 114–145.

Solomon, J., & Nemirovsky, R. (in press). Mathematical conversations [Videopaper]. *Journal of Research in Mathematics Education.*

Turner, J. S. (2000). *The extended organism.* Cambridge, MA: Harvard University Press.

Wittgenstein, L. (1953). *Philosophical investigations.* New York: Macmillan.

Part III

TEACHING FOR UNDERSTANDING

Chapter 9

Changing Teachers' Professional Work in Mathematics: One School's Journey

Megan Franke
University of California–Los Angeles

Elham Kazemi
University of Washington

Jeff Shih
University of Nevada–Las Vegas

Stephanie Biagetti
California State University–Fullerton

Daniel Battey
University of California–Los Angeles

The story of learning at Crestview Elementary is difficult to characterize in a linear, tidy story. For this reason, we chose to describe this school's difficulties, movements, challenges, and successes, not through traditional data analyses, but through a series of defining moments. These moments made explicit the defining and redefining of knowledge, skills, and stance of the school community as well as the evolution of identities. Our work with the teachers and staff taught us that some of the best practices often adhered to, such as that good professional development takes time, might mean something different given the context of the work, the people involved, and the focus of the professional development work. In what follows, we describe the focus of this work, the building of connections between the professional development and the classroom, the travel of professional development beyond the classroom, and the resulting evolution at Crestview in light of these defining moments.

FOCUSING PROFESSIONAL DEVELOPMENT

Our work with Crestview Elementary was guided by a theory of learning, a set of principles that emerged from that theory, and existing research on professional development (see Franke & Kazemi, in press). We explicitly focused on the development of students' mathematical thinking and, using examples of student work as the basis for discussion and interaction, created opportunities for teachers to develop communities of practice around teacher inquiry. Our goal in this section is to make explicit the focus of this work, both in substance and structure, and the theories that supported it.

Conceptualizing Teacher Learning

Our work is driven by our conceptualization of teacher learning, which draws in particular on situated theories of learning. Within a situated perspective, understanding learning as it emerges in activity is paramount (Greeno & MMAP, 1998). We draw on Lave's (1991) description of learning, thinking, and knowing as "relations among people engaged in activity *in, with, and arising from the socially and culturally structured world*" (p. 67, emphasis in original). In our work with teachers generally, we think about the ways teachers engage in activity, the ways we can support their work within a community of practice, and the roles that tools, artifacts, and participation structures have in teachers' evolving practices (Wertsch, 1998). Importantly, we also recognize that we can understand teachers' learning only as it relates to those with whom they engage and to the context in which that engagement occurs.

 We view teachers' collective inquiry within schools and within their classrooms as forming overlapping communities of practice (Lave & Wenger, 1991; Wenger, 1998) that continually define the work of the community as well as the resources that help teachers negotiate the meaning of their work (Wenger, 1999). Lave and Wenger (1991) and Rogoff (1994, 1997) described learning as shifts in participants' roles and identities linked to new knowledge and skill, with roles varying not only across time but also within individuals over time (Rogoff, 1994). Lave (1996) noted that "crafting identities is a *social* process, and becoming more knowledgeably skilled is an aspect of participation in social practice, who you are becoming shapes crucially and fundamentally what you know" (p. 57).

In framing this shift, Wertsch (1991, 1992) highlighted learning as occurring through mediated action between the individual and the sociocultural setting, with cultural tools affording or constraining learning or change. Wertsch's conceptualization highlights the critical nature of the tools and artifacts that serve to mediate the action between teachers and school (and other) cultures within communities of practice. A number of mathematics educators (see, e.g., Boaler, 2001; Cobb, 1999; Cobb, McClain, de Silva Lamberg, & Dean, 2003; Stein & Brown, 1997) also draw on a similar situated perspective, describing teacher learning as interactively constituted in complex learning environments backgrounded by mathematical content and tools.

Conceptualizing Professional Development

Creating a Community of Learners. We resonate conceptually with creating school learning communities (Sarason, 1996) in which teachers see themselves as intellectuals (Giroux, 1988; Little, 1999) and recognize the importance of teachers learning through activities embedded in their everyday work (see also Fullan, 1991; Lieberman & Miller, 1990; McLaughlin & Talbert, 1993; Secada & Adajian, 1997; Tharp & Gallimore, 1988). Wenger (1998) described the need to find

> inventive ways of engaging students in meaningful practices, of providing access to resources that enhance their participation, of opening their horizons so they can put themselves on learning trajectories they can identify with, and of involving them in actions, discussions, and reflections that make a difference to the communities they value. (p. 10)

Our intent was to create space, as Wenger (1998) described, for teachers to share, challenge, and create ideas about the development of children's mathematical thinking, and in which teachers could shape their identities and take on a "stance," defined as the "ways we stand, the ways we see, and the lenses we see through" (Cochran-Smith & Lytle, 1999, p. 49). Specifically, our goal included supporting teacher workgroup communities in which teachers came together to share and make sense of their students' mathematical thinking and make public their private acts of teaching. We wanted to provide teachers the forum to develop relationships and create a community of practice that included the interactions and identities developed in their classrooms.

Engaging in Inquiry. Many researchers discuss the critical need to understand *what* teachers engage in inquiry about and how they engage in that inquiry (Grossman, Wineburg, & Woolworth, 2000; Lehrer & Schauble,

1998; Little, 1999; Lord, 1994; Richardson, 1990, 1994a; Schifter, 1997; Warren & Rosebery, 1995). Cochran-Smith and Lytle (1999) noted that "teachers across the life span play a central and critical role in generating knowledge of practice by making their classrooms and schools sites for inquiry, connecting their work in schools to larger issues, and taking a critical perspective on the theory and research of others" (p. 273).

We focused on teacher inquiry into the specifics and evolution of students' mathematical thinking. In our initial work, we drew on Cognitively Guided Instruction (CGI; see Carpenter, Fennema, & Franke, 1996; Fennema et al., 1996), one of a number of projects using student learning trajectories (see, e.g., Barnett & Sather, 1992; Borko & Putnam, 1996; Brown & Campione, 1996) to engage teachers in the study of their students' mathematical work. We did not provide teachers with specific details about CGI learning trajectories, but used them to guide our interactions and discussions with teachers.

Identifying Artifacts.   Wenger (1998) argued that tools or artifacts (in our case, student work) can provide a focus for the negotiation of meaning within a community and can foster understanding, a process he termed *reification*. The artifacts support the development of shared language as well as interaction that can support the development of new relationships. We see the use of artifacts (in this case, mathematical problems) as a way to focus the conversation, bring teachers' rich experiences and histories into play, and create connections between professional development and teachers' work. The monthly workgroup problem, which teachers posed to their students, was designed to permit a range of entry points to accommodate different mathematical understandings. (As the sessions evolved, the teachers participated in choosing the tasks.) During the sessions, the teachers shared their students' work, created a group list of elicited strategies (ranked the list according to sophistication of mathematical understanding), and talked about implications. All conversation revolved around the particular problem posed and included discussion of the students' mathematical thinking in their solving of the problem, the mathematics in the problem itself, the details and usefulness of particular strategies, and ways to move students' mathematical thinking forward. (At each meeting, we also asked the teachers to respond in writing to reflection questions related to their student work and classroom practices and collected their student work.)

Developing Relationships.   Our work with Crestview Elementary began as a study of professional development and evolved into a study of school change. After working at Crestview informally for 2 years, we proposed to the staff a systematic study of mathematics professional development focused on understanding the development of students' mathematical thinking (Carpenter et al., 1996). We engaged interested teachers as well as those encouraged to participate by the principal (all together more than two thirds of the staff). We wanted teachers to work together to create trajectories of student thinking based on their interactions with their own students.

The workgroups included Grades K–5 teachers from one of four tracks (approximately 8–12 teachers per workgroup) as well as the principal, other administrators, and staff. We saw the formation of relationships, both within and outside of the workgroup, as critical to our learning, to the development of communities of learners, and to the potential for this work to become self-sustaining. We wanted the teachers to see us participating with them in learning about the development of students' mathematical thinking and about ways to incorporate that knowledge into their classroom practice. We wanted them to see from the start that they were experts both about their own students and about ways to support the learning of their students.

To this end, the workgroups began and ended with informal conversations. In between meetings, we visited their classrooms informally to observe and document students' thinking. We wanted the teachers to view us not as having answers, but as having questions and being engaged in continual learning about their students' mathematical thinking. We also wanted to become part of the broader school community, to which end we spent time in the lunchroom and on the play yard, wandered through the school, met regularly with the principal, and talked with the administrators and staff. We shared our histories and our personal lives because we wanted to create an exchange that would provide opportunity to learn about different aspects of teachers' work and its impact on their lives.

At the end of our first year together, we brought to a close our specific study of professional development, but the teachers wanted us to continue. At the teachers' request, we kept the focus on student work, but changed the workgroup meetings to include a rotation between grade-level and track-level meetings. More teachers opted to participate. We continued the same workgroup structure and continued to lead the sessions. At the end of the second year, together with the principal, we decided to invite teachers to lead the workgroups. At the same time, the principal found money to hire one of the teachers as a half-time mathematics specialist. In the third year, the mathematics team facilitators (MTFs; their name for themselves) led the workgroups. We provided support to the MTFs, as did the principal and the mathematics specialist. The workgroups currently continue, 4 years after we moved out of our role as facilitators.

CONNECTING PROFESSIONAL DEVELOPMENT
TO THE CLASSROOM

Initially, teachers saw the workgroups as disparate meetings they attended, not something connected to their teaching practice or their classroom work—even though the discussions focused on the work of their own students. Over time, the teachers began to see the workgroups as part of their practice although the individual evolution of teacher perspective differed. In what follows, we focus on particular moments of shift.

Seeing the Connections

Halfway through the first year, the teachers posed their students a series of three computation problems involving addition and subtraction, with regrouping. Up to this point, all of the monthly problems had been word problems. As the workgroup meeting began, Ms. P came in the door talking excitedly about her experience posing the problems. We had not officially started the meeting, and people continued wandering in, but she could not wait to talk about what had happened. She told the group, "All this time I kept doing my mini-lessons." In these mini-lessons, which she taught before posing the workshop problems, she showed the students how to solve a problem like the workgroup problem. She then walked the students through the problem step-by-step to make sure they understood. At that point, she posed the workgroup problem. For no preplanned reason, this time she had simply posed the problem to her students without preteaching. And they had solved the problem—using many more strategies. She told the workgroup:

> I have been doing all these mini-lessons, and I didn't need to. They can solve the problems without them. I can't believe I have been doing this all this time. I realize now, I guess I just, that they all would just use the strategy I showed in the mini-lesson, so I always got the same strategies, never got a range of strategies.

The teachers talked about Ms. P's experience for over 20 minutes. They asked her questions and wondered about their own classrooms. The teachers brought out their student work and looked across teachers to see who was getting a range of strategies. They asked each other how they led up to posing the problem and wondered whether and in what ways that might influence the strategies they were seeing. Some teachers reported that their students needed the preteaching; others wanted to see what would happen if they didn't preteach.

Although this workgroup moment raised many issues, what became clear to all was that after that interchange the workgroups changed. Teachers more frequently discussed how their actions might influence students' strategies. They talked more about experimentation: "What happens if we do this? Let's try it and see." Teachers began to sense that students had strategies they didn't know about—and they wanted to find out about them. Although teachers took different stances on students' mathematical thinking and ways to make use of that reasoning, the discussion had begun. Practices were being made explicit, and student work was used as evidence. The teachers saw a connection between their actions in the classroom and their learning in the workgroup.

Engaging in Argument

Although we had designed the workgroups to allow space for teachers to question themselves and their student work as well as to challenge each other, we were almost surprised when teachers began to engage in argument around the learning and teaching of mathematics. The way argument evolved differed across workgroups, but here we discuss what we consider to be the defining moment in one workgroup's evolution.

During the sixth workgroup meeting, the teachers challenged one another's long-standing ideas about the standard algorithm. The teachers pushed each other and listened, but also looked hard at student work for evidence to support their positions. The conversation was long and involved. The exchange we provide here also illustrates teachers' use of student thinking to support their arguments, specifically a list of strategies the group had generated earlier from their students' solutions:

Facilitator:	What do you think is the most sophisticated strategy that you saw?
Mr. M:	Probably using the algorithm.
Facilitator:	The algorithm? Do you all agree about that?
Teachers:	No.
Ms. N:	They don't know what they're doing [when they do the algorithm].
Ms. K:	And you don't have to think about the problem when you're doing it. Like you have to think about 40 as a unit in other strategies.
Mr. M:	But you have to hold the value when you multiply.
Ms. K:	That's why I think it's—
Mr. M:	That's sophisticated in itself.
Ms. K:	But I think that's a memorized rule. I don't think kids think, "Oh, I'm holding the 1s place here."
Ms. N:	It would be interesting to see if they could explain why they're multiplying and what it means.

Mr. Q:	Yeah, I think it is. For a lot of kids, it's a process that they learn.
Mr. M:	I have a question. Why do they separate it into 40s and 7s?
Ms. K:	Because it's really adding 40s. I mean you're really adding—
Mr. M:	That's what I mean. It's really adding. As opposed to, I consider multiplication just a higher level.... As long as you know how to add double-digit numbers, that's not very complicated. But the idea of conceptualizing the fact that, you know, groups of—
Ms. K:	That's why I think the thought process involved while they're doing multiplication is more they're doing it out of memorization. They're just doing, like, a procedure that they've memorized. They're not thinking about what the numbers mean and what they represent whereas the one where they're adding the 40s, they're making sense of the problem. And they're actually more likely to be right.
Ms. Z:	Yeah, that's what Ms. R just said.
Ms. N:	I think that, like on number 7 [on the list of strategies], for example, like Vincent, he solved it by doing the algorithm, but he explained, "Mrs. Mack used 705 dollars to buy animal cookies. Each box cost 47 cents, and she bought 15 boxes, so I multiplied 15 and 47." So he was able to explain why he did it.
Mr. M:	One thing why I said the algorithm is important for me is that later on when you get—I don't know if this is on the off chance—but if you get into physics and chemistry, you start memorizing constants and stuff.
Mr. Q:	I think [strategies] #3 and #4 are a pretty clear differentiation in place value, which is sort of the critical thing. So they [the students using these strategies] are understanding, the person who added up rows of 4 and knew it was 10s. And added up rows of 7 and knew it was the units, that kind of blows me away.
Ms. N:	They were able to keep track of it.
Mr. Q:	Also the people who knew how to do the lines and circles [pointing out the direct modeling by 10s strategy].... That shows that they've really got the concept of groups.

Not all of the teachers in the conversation agreed about the most sophisticated strategy, nor did they all use specifics of student thinking to marshal their evidence. The details of student thinking, however, allowed the teachers to engage in a discussion that was mathematical, detailed in students' strategy use, and thorough in examining the issues involved in using the standard algorithm. Even more importantly though, this passage shows that teachers could take the lead in marshalling an argument. Initially, teachers did not want to disagree with each other or push at each other's thinking. When Ms. N explicitly disagreed, however, the conversation opened up. The teachers continued discussing with each other, working through disagreements, and pressing at long-standing contentious issues. The teachers challenged one another's opinions, drawing on evidence from their practice, and listened and responded to arguments of their colleagues.

PROFESSIONAL DEVELOPMENT: TRAVELING BEYOND THE CLASSROOM

Reconstructing Supervision

The principal (Ms. J) and all of the administrative staff participated in one of the workgroups as learners. During our first year of work together, Ms. J adapted her practice as principal in a way that exemplified the role she played in supporting innovation: She decided to focus on student work in "evaluating" her teachers. Rather than observing in their classrooms and documenting teachers' pedagogical practice, Ms. J asked the teachers to come to discuss their students' mathematical work with her. When asked why she had made this change, Ms. J replied:

> I was dissatisfied with observation because I found that anybody could do anything for an hour, no matter what they were doing on a daily basis. And little of what I saw, or we talked about, ever got to how the students were doing ... I use the student work as a way to back into the teaching. It has to start with the students. What are they understanding, or not understanding, and then what are we going to do? What teaching strategies are you going to use to address that? ... The biggest thing that focusing on student work accomplishes is that it tunes us into student work—it sets the tone that we use student work to talk about teaching. All decisions in the classroom are based on what you see in the student work, not how the lesson goes.

So the teachers brought Ms. J examples of their students' mathematical work three times during the year. Ms. J described the practice:

> We start at the beginning of the year by having the teacher identify her goals in the content area. The teacher then brings student work that she thinks reflects her students' progress toward those goals. The teachers bring work of those that they think are doing well and those they are concerned about. I ask them what the students who are doing well are able to do and have them show me the work. We talk about what the students they are concerned about are doing. We look at the strategies the teachers are using with the students that are doing well and the strategies they are us-

FRANKE ET AL.

ing with the students they are concerned about, and then we talk about
what they think they should be doing next. I then use this as a basis to visit
their classroom and look at the work of specific students.

During the fall, we sat in on some of these meetings. They typically followed the same pattern. For example, Ms. D brought a class set of student
solutions to a workgroup problem. Ms. J asked her to discuss students'
thinking and what their solutions showed about their understanding of the
mathematics involved. Ms. D picked out the work of a few children, describing what they knew using phrases like, "she could do it easily," "he seemed
to struggle," and "he needs work with place value." Ms. J then asked her,
"What do you think they know about 10s and using 10s?" When pushed to
detail student thinking, Ms. D reflected on what student work told her (or
did not tell her) about a student's understanding. After a while, Ms. J asked,
"What would you like the students to know?" In her reply, Ms. D discussed
the kinds of problems she would try next, the student thinking she expected
to see, and what she would do next with particular students.

Ms. J played a role in supporting teachers' understanding and use of students' mathematical thinking:

> Well, at the beginning, I guess, I saw some more limited student work,
> more worksheet type of work ... I ask the teachers, so what is this telling
> you? What are you trying, what strategies are you using to support this stu-
> dent? How are we going to see change? So if I am teaching this way, what
> am I learning?

She felt that no matter what teachers brought in terms of student work, or
their perspectives on the teaching and learning of mathematics, there was a
conversation that could be had comfortably, during which she could press
them about students' thinking and what it meant for a teacher's own practice:

> It is hard for a teacher to sit with you one-on-one and not talk about how
> he can help a child. I found that this was a way that I could push them to
> see what students could do and I could ask them, "So what are you going to
> do to help this child?" The teacher would have to say something, and I
> could ask for specifics. They could not leave it at saying the child wasn't ca-
> pable or just can't do it. They had to leave with some plan that involved
> what they would do to help.

Ms. J showed the teachers through her work with them (and in a very visible way) that she valued knowing about student thinking and wanted all the
teachers to be able to talk about it regardless of their classroom decisions.
Ms. J supported the teachers in regarding themselves as people knowledgeable about student thinking.

Supervision is not a place where one expects innovation. Supervision is often primarily about power, hierarchy, and ritual. Here, Ms. J chose to share supervisory power. The teacher was the expert about how his or her students solved problems, the teacher chose the student work to focus the conversation, and the teacher and the principal engaged together in discussion that supported the work of the school.

These principal–teacher interactions moved the focus on student work outside the workgroup. The teachers used the workgroup language and tools in the meetings with the principal. They became knowledgeable about their students' reasoning and used what they knew in their meetings with the principal. Workgroup meetings, classroom practice, and evaluation were all linked by a focus on student thinking.

Developing School-Wide Conversation

During the second year of the project, one second-grade teacher, Mr. A, decided that because his students had developed such little understanding of place value the previous year, he needed to do something differently. He felt that his students moved too quickly to the algorithm, with little understanding of place value. He adamantly said he would not let that happen again and told himself and us that no student would leave his classroom without understanding the relationship between 1s, 10s, and 100s. He used that as the focus of his attention and his mathematics curriculum for the entire year. He covered many other mathematical topics, but he related those (when he could) to understanding the relationship between 1s, 10s, and 100s. He used multiplication and division, measurement, money, and patterning to discuss these particular relationships. At the end of the year, he felt that this focus had paid off. (Our data support this.) He felt his students ended the year understanding these relationships and that none of them used the standard algorithm without understanding it.

Ms. L also created and maintained a focus (equivalence and fractions) in her work with her fourth and fifth graders. When these teachers discussed their ongoing experimentation in the workgroup meetings, others became intrigued, although not all their colleagues could create or sustain such a focus. But as the end of the year approached, the teachers raised this as a possibility for a school articulation meeting.

Toward the end of the year, one faculty meeting was devoted to identifying, by grade level, key mathematical ideas that students needed to understand. This conversation proved to be another defining moment. The teachers who had participated in the workgroups had a language they could readily use to explain their ideas. Using evidence from their students' learning, they argued about key ideas and ways to frame the issues. Although the workgroup teachers did not agree on everything, they did not hesitate to put forth an opinion, and they were willing to engage in argument. Workgroup teachers, in effect, extended the workgroup norms to this whole-school context. Teachers discussed patterns, identifying those they wanted students to understand. They focused on detail and pushed the conversations beyond general labels and terms like patterning and place value, which often carry multiple meanings: Do we talk about place value? Or do we talk about understanding relationships between 1s, 10s, and 100s? When we say place value, do we mean identifying the 10s place? They talked about what it meant to understand equivalence and discussed ways to extend mathematical ideas. Some of the familiar workgroup discussions (e.g., about where the standard regrouping algorithm fit in the curriculum) were brought into this conversation. They discussed how ideas like learning number facts fit in with the key ideas they had identified. A few teachers who had not been participants were pushed hard to explain the rationale for their perspective. They were challenged and questioned in ways that showed a particular expectation for what constituted an acceptable rationale. These teachers sustained their conversation until they had tentative grade-level frameworks that they wanted to use and evaluate during the following school year.

Although the workgroups had focused on small groups of teachers engaging in discussions about the learning and teaching of mathematics, the effects of the workgroups continued to emerge in the whole-school meetings and conversations: The norms for participation in the broader school community were being reestablished and renegotiated.

Influencing District-Level Policy

The Crestview teachers continued to develop knowledge and skills around the development of students' mathematical thinking and identities. As teachers realized that they could argue about what to do in teaching mathematics, they not only began to talk with each other more, but they also joined in broader discussions outside the school context. During second year, Ms. V, Ms. N, and Ms. K (three of the workgroup teachers) volunteered to serve on the district-wide curriculum committee—for a particular reason:

> We heard Schmidt, the guy from the [Third International Mathematics and Science Study] study, talk about the problems with math education in America on our opening day, and [we] came to the committee the next week asking how we could make changes in accordance with the findings he presented. I and a few others were determined not to let such an opportunity for change pass us by. We felt that the district would support this radical change ... after all, they invited Schmidt to speak to us! We realized that we were trashing three years' work by the committee but knew that it was the only way we could fix the problem with math achievement.

Ms. V, Ms. N, and Ms. K joined the committee feeling that, "We have something to say. We want to make sure that we can keep doing what we think is the best thing to do in mathematics." But they were also hesitant. They recognized that some of their ideas and opinions were quite different from those of their district colleagues. They worried about how their notions of teaching mathematics would be perceived and whether these notions would be valued across the other schools in the district.

When the district rolled out its new mathematics standards, however, the Crestview teachers were struck by the long list of skills specified for each grade level. They were equally surprised that these particular standards were introduced on the same day that Dr. Schmidt had described a common problematic approach to mathematics education in U.S. schools: the teaching of a long list of skills rather than a deep exploration of a few areas of mathematics. At the first meeting of the mathematics curriculum committee meeting, Crestview teachers proposed that the district standards be rethought. Surprisingly (to the teachers, to the principal, and to us), the superintendent and the rest of the committee agreed that the issue should at least be addressed.

The Crestview teachers on the curriculum committee took the lead. They told us, "This is it. We can do this. We have to—for us and the kids." They were not worried about being able to come up with a new set of standards. They had already decided that the separate list of skills needed to be regrouped under more conceptual headings for each grade level. Instead, they worried about creating a document that would be acceptable to their district colleagues. They began by grouping the standards under the grade-level conceptual headings Crestview had recently identified. They shared their groupings with their colleagues, with the principal, and with us. They then made adjustments, practiced presentation of their standards, and took them back to the committee. These teachers ended up convincing not only the curriculum committee to adopt this form of the standards, but also convinced the superintendent and the school board of its value.

Although the process was a challenging one, the teachers were excited about their ability to argue their case. They had evidence from their classrooms and research they themselves had done in their classrooms that suggested that their new version of the standards would be more helpful to teachers and a better fit for the development of their students' mathematical thinking. The district later adopted the Crestview teachers' adaptation of the standards. The teachers had understood that they had the knowledge to make their voices heard and had realized that the other teachers on the committee were unsure where to go with the standards. They came to realize that they had been effective in changing the district standards because they had known exactly what they wanted and had been able to articulate and to support their reasoning.

These Crestview teachers not only took on a new role and set of responsibilities, but they also did so within a very political climate. They had to take on political arguments about what was and was not good for students in mathematics. They had to figure out how to get their points across without sounding dogmatic. Often the teachers came back to Crestview after a meeting and tried out ways of presenting their arguments. They took on the stance of expert in a way that they did not have to or want to within their school, but in working with the district committee, they saw themselves as representing Crestview and what the teachers at Crestview considered important for students to learn about mathematics. These teachers were passionate about their district-level work because they felt they were addressing a particular issue that had an impact on student achievement and because they believed they had expertise to offer.

Generating Leadership

At the end of the second year, the research team planned to move out of the workgroup leadership role and hoped that the school would in some way be able to continue the work. We recognized, however, that this could not happen without some conversation and planning. The principal worked out a way to release one teacher half time to provide leadership in mathematics, but quickly realized that the situation would devolve into the traditional model, with the lead teacher becoming the mathematics specialist. The principal came to us with the idea of recruiting a group of teachers within the school interested in taking on leadership roles. We discussed who these teachers might be and why. She decided it would be best to see who was interested and let everyone have the opportunity. So she sent a memo inviting any interested teachers to come to talk with her. She also specifically reached out to a few she thought would provide a broad base of support. Those interested (8–10 teachers from different workgroups) met and discussed what their role might look like and what they might work to accomplish. One member of the research team offered to meet monthly with the leadership team to support them in their work.

The leadership team decided there was enough school support to continue, so each teacher (or pair of teachers) from the leadership team became responsible for facilitating a workgroup. The leadership group talked at length at several meetings about what to call themselves. They decided on mathematics team facilitators (MTFs) because they did not want people to see them as more expert, just willing to help facilitate the group's learning. The leadership group discussed possible strategies, created a plan to cover the next few meetings, and began their workgroups. After the second workgroup meeting, the MTFs reported in the leadership meeting that they were disappointed in how their workgroups seemed to be going. They felt they were "not engaging the teachers in ways that are productive." Each MTF described the workgroup meeting he or she had facilitated, and the group as a whole followed up with probing questions. At the end of the meeting, the MTFs concluded that they had "moved away from what is most important—student work." Almost all of them realized that they had spent a significant amount of time describing their own classrooms and the pedagogy they used to engage students in mathematics. In reflecting on the discussions they had led, they realized that their sharing in the workgroups had not been focused on *student work*.

The MTFs reminisced about their own 2 years of workgroup experiences. They discussed how powerful talking about their own students' work had been, especially connected to pedagogy. The MTFs decided they needed to return to a focus on student work and student thinking. They returned to the key ideas teachers had set for their grade levels and developed problems the teachers in the new workgroups could pose to their students. The MTFs felt that this process would allow teachers to see what they had accomplished through its effect on their students' work. This shift in workgroup focus marked a defining moment, and although the structure of the workgroups continued to evolve, the MTFs kept student work central to the conversation.

The struggle around planning the workgroups was not unusual or particularly surprising. The MTFs were defining and developing for themselves new roles, new sets of responsibilities, and new identities as leaders in the school context. The initial workgroups had drawn on more established interaction patterns among the teachers, and the MTFs, questioned about how they taught mathematics, had spent a significant amount of the time talking about their classroom practice—even when they had not intended to do so. The workgroups had to adjust to the new roles of the these colleagues and to the our absence. But it was the MTFs who recognized the workgroup patterns and created ways to develop a set of norms more consistent with school goals. In their discussion, they had in effect gone back to the artifacts from the earlier workgroup meetings, which they saw as in line with their goals. At their next workgroup meetings, the MTFs returned to the old workgroup structure: Teachers brought student work for a problem that the MTFs had distributed, they answered reflection questions, and they talked about the student thinking. The workgroups continued in this pattern, and the problems they posed became tied to quarterly benchmark tasks for each grade level.

As the year progressed, the MTFs realized that the leadership meetings extended their own learning. They began to view their participation in the leadership team as a learning opportunity, one they thought all teachers needed. At the end of a year of their leadership, the original MTFs proposed that the leadership rotate so that all teachers could take on the role at some point.

EVOLVING INTO A LEARNING COMMUNITY

Teachers at Crestview did not suddenly take on a stance of inquiry, nor did participants engage in the same way or move toward full participation at the same time. The defining moments we have described here illustrate the complexity involved in the ways learning occurred and community developed. Although we find ourselves wanting continually to construct and simplify lists that represent how the learning occurred at Crestview, these moments remind us that this story of learning is anything but simple.

Central to each of these instances was a struggle around making sense of practice. Participants examined their practice both for what worked, and what did not, based on evidence from student work. Teachers found ways of evaluating and justifying practice beyond their sense that something was a "good idea" or was "working." The participants also perceived that their struggle was ongoing. The teachers and administrators became better at recognizing what might not be working and more willing to share their struggles with their colleagues. The struggle occurred through the workgroups, in teachers' engagement with the principal, in their collaborations, in their work as a school community, in their work with the district, and so on. Importantly, teachers were willing to continue to struggle and engage in sense-making related to the focus of their engagement, the evolving and established structures, and their developing identities.

Although we cannot separate these elements from each other, we can highlight how they played out. We saw emerging teacher insight, like that of Ms. P, who came to challenge her own practice in relation to what she was learning about the development of children's thinking. We also saw teachers like Mr. A and Ms. L, who engaged in inquiry as they taught, using what they knew about the development of students' mathematical thinking to create a plan for working with their students and to continually evaluate how that plan progressed. Teachers engaged in inquiry with the principal as they shared student work over the year and as a community as they determined key mathematical ideas for each grade level. As they engaged in inquiry across multiple settings with different participants, the focus on students' mathematical thinking remained consistent. Although not all teachers and administrators engaged in inquiry (or engaged consistently), a school-wide culture of inquiry based around student thinking developed over time.

As teachers began to see the workgroup focus on students' mathematical thinking as something related to their interactions with their students in their classrooms, their stance toward inquiry changed. Participating in inquiry became about continuing their own learning, rather than about helping the researchers out or doing what they thought their principal wanted. There was a growing recognition and explicit acceptance that teachers' work occurred not only in the classroom, but also as they collaborated with other teachers, the principal, researchers. Inquiry around student thinking was explicitly tied to different aspects of their practice—interacting with students, planning lessons, creating district math standards, being evaluated, and so on.

Before the research team and teachers had begun their collaboration, teachers' inquiry around their practice was conducted only in isolation, as individual teachers working in their classrooms. We began our focused inquiry together in the workgroup, and although the workgroups evolved and the teachers began more often to guide the work in those groups, inquiry was not limited to the context of the workgroup. The focus and structures of inquiry afforded opportunities for leadership to become shared by a wide range of participants. In some sense, each teacher became expert about his or her own students' thinking. They each took on leadership roles as they shared this expertise with the group. Our role as experts on students' mathematical thinking also fostered their leadership in that we respected teachers as the natural experts about pedagogy. The teachers increasingly looked to one another for support in making use of the students' thinking in their classrooms. As their knowledge and skills developed, the teachers began taking on leadership roles both within and outside of the workgroups.

The shifts in the teachers' roles were reflected in their developing identities around the teaching and learning of mathematics and as leaders. The teachers began as participants unsure of any shared goal, but as they entered more into communal engagement in the "cause," they began to see themselves as different from other teachers in other schools in their thinking about, and approach to, teaching mathematics. Many told us and each other that they could never go back, and they were not sure they could teach anywhere else. They continued to look for new challenges, and those challenges often took them into more political arenas. Some teachers were less sure about the shift they perceived occurring in the school. Some moved to other schools; others began paying attention to what was happening from the sidelines.

The research already completed around communities of practice, school change, and professional development provides a great deal of insight into understanding the evolution at Crestview and in many ways supports our contention about the complexity of Crestview's development. The changes the teachers made in thinking about the "what" of inquiry, extending the focus of inquiry into other aspects of their practice, including their classrooms, defining a stance, and creating identity are critical to understanding what occurred, and our analysis of those changes underscores the complexity of professional development. In retrospect, we also recognize how leadership and the different forms of leadership played a role in the developing community.

Particularly, we noticed what can happen when leadership decisions align with the professional development, opportunities for developing leadership allow leaders to be learners, and all members of the community are encouraged in different ways to take on leadership roles (Lord, 1994; Spillane & Thompson, 1997). Our developing understanding of the evolution at Crestview provides insight into the ways the work of these scholars as a whole can be seen as a frame for understanding school learning.

CONCLUSIONS

That Crestview remained a thriving community of inquiry long after we left suggests that no professional development list of best practices, even those that include engaging teachers in inquiry, focusing on students' mathematical thinking, and attending to the developing community of practice, alone provides enough guidance or detail to another professional developer in designing and implementing a program. Any list of best practices can be construed and detailed in ways consistent neither with the theories we used to drive our practice nor with the local situation; in either case, such a list can lead to less than successful professional development. That said, we do believe that our work helps to create knowledge about professional development.

We know that what happened during these 3 years at Crestview was about more than teachers gaining knowledge about the development of children's thinking—although we could argue that growth in teachers' knowledge was certainly at the core. What happened at Crestview had to do with teachers inquiring consistently and specifically about their practice *with their colleagues*. Teachers began to see themselves as a part of a community where they worked together on their practice—not that everyone had to agree—but that it was important to find ways to talk with each other about disagreement. What happened at Crestview was that the teachers developed identities about the teaching of mathematics and came to see themselves as learners, as leaders, and as researchers. They came to realize that they could put forth a reasoned argument supported by evidence they themselves collected about why a particular approach was, or was not, a good idea. They were able to go beyond simply agreeing or disagreeing. What happened at Crestview had a core focus that gradually permeated many school practices, required seeing and supporting informal and formal leadership at all levels (including that taken by individual teachers in their workgroup conversations), and evolved and continued to evolve *because of the people and their relationships*.

What happened at Crestview did not happen smoothly, nor has it finished happening. What we hope is that this story and others like it will challenge professional developers to further engage themselves and us in ways that will continue to enrich the theories and practices that drive all of our work.

AUTHOR NOTE

The research reported in this paper was supported in part by a grant from the Department of Education Office of Educational Research and Improvement to the National Center for Improving Student Learning and Achievement in Mathematics and Science (R305A60007-98). The opinions expressed in this paper do not necessarily reflect the position, policy, or endorsement of the funding agencies. We thank the participating teachers and administrators and gratefully acknowledge the contributions of the SSGC research group. An earlier version of this paper was presented at the annual meeting of the American Educational Research Association, 2000.

REFERENCES

Barnett, C. S., & Sather, S. (1992). *Using case discussions to promote changes in beliefs among mathematics teachers*. Paper presented at the annual meeting of the American Educational Research Association, San Francisco, CA.

Boaler, J. (2001). *Opening the dimensions of mathematical capability: The development of knowledge, practice, and identity in mathematics classrooms*. Paper presented at the meeting of the North American Chapter of Psychology of Mathematics Education, Snowbird, UT.

Borko, H., & Putnam, R. (1996). Learning to teach. In D. Berliner & R. Calfee (Eds.), *Handbook of educational psychology* (pp. 673–708). New York, NY: Macmillan.

Brown, A., & Campione, J. (1996). Psychological theory and the design of innovative learning environments: On procedures, principles, and systems. In L. Schauble & R. Glaser (Eds.), *Innovations in learning* (pp. 289–326). Mahwah, NJ: Lawrence Erlbaum Associates.

Carpenter, T. P., Fennema, E., & Franke, M. L. (1996). Cognitively guided instruction: A knowledge base for reform in primary mathematics instruction. *Elementary School Journal, 97*(1), 1–20.

Cobb, P. (1999). Individual and collective mathematical development: The case of statistical data analysis. *Mathematical Thinking and Learning, 1,* 5–43.

Cobb, P., McClain, K., de Silva Lamberg, D., & Dean, C. (2003). Situating teachers' instructional practices in the institutional setting of the school and district. *Educational Researcher, 32*(6), 13–24.

Cochran-Smith, M., & Lytle, S. L. (1999). Relationships of knowledge and practice: Teacher learning in communities. In A. Iran-Nejad & C. D. Pearson (Eds.), *Review of research in education* (Vol. 24, pp. 249–305). Washington, DC: American Educational Research Association.

Fennema, E., Carpenter, T. P., Franke, M. L., Levi, L., Jacobs, V., & Empson, S. (1996). A longitudinal study of learning to use children's thinking in mathematics instruction. *Journal for Research in Mathematics Education, 27*, 403–434.

Franke, M. L., & Kazemi, E. (in press). Teaching as learning within a community of practice: Characterizing generative growth. In. T. Wood & B. Nelson (Eds.), *Beyond classical pedagogy in elementary mathematics: The nature of facilitative teaching*. Mahwah, NJ: Lawrence Erlbaum Associates.

Fullan, M. (1991). *The new meaning of educational change* (2nd ed.). London: Cassell.

Giroux, H. A. (1988). *Teachers as intellectuals: Towards a critical pedagogy of learning.* Granby, MA: Bergin & Garvey.

Grossman, P., & Wineburg, S., & Woolworth, G. (2000). *What makes teacher community different from a gathering of teachers?* Paper published by the Center for the Study of Teaching and Policy, University of Washington.

Lave, J. (1996). Teaching, as learning, in practice. *Mind, Culture, and Activity, 3,* 149–164.

Lave, J., & Wenger, E. (1991). *Situated learning: Legitimate peripheral participation.* Cambridge, England: Cambridge University Press.

Lehrer, R., & Schauble, L. (1998). *Modeling in mathematics and science.* Madison: National Center for Improving Student Learning and Achievement in Mathematics and Science, Wisconsin Center for Educational Research.

Lieberman, A., & Miller, L. (1990). Teacher development in professional practice schools. *Teachers College Record, 92*(1), 105–122.

Little, J. W. (1999). Organizing schools for teacher learning. In L. Darling-Hammond & G. Sykes (Eds.), *Teaching as the learning profession: Handbook of policy and practice* (pp. 233–262). San Francisco: Jossey-Bass.

Lord, B. (1994). Teachers' professional development: Critical colleagueship and the roles of professional communities. In N. Cobb (Ed.), *The future of education: Perspectives on national standards in America* (pp. 175–204). New York: The College Board.

McLaughlin, M. W., & Talbert, J. E. (1993). *Contexts that matter for teaching and learning: Strategic opportunities for meeting the nation's educational goals.* Stanford, CA: Center for Research on the Context of Secondary School Teaching, Stanford University.

Richardson, V. (1990). Significant and worthwhile change in teaching practice. *Educational Researcher, 19*(7), 10–18.

Richardson, V. (1994a). Conducting research on practice. *Educational Researcher, 23*(5), 5–10.

Richardson, V. (1994b). *Teacher change and the staff development process: A case in reading instruction.* New York: Teachers College Press.

Rogoff, B. (1994). Developing understanding of the idea of communities of learners. *Mind, Culture, and Activity, 1,* 209–229.

Rogoff, B. (1997). Evaluating development in the process of participation: Theory, methods, and practice building on each other. In E. Amsel & K. A. Renninger (Eds.), *Change and development: Issues of theory, method, and application* (pp. 265–285). Mahwah, NJ: Lawrence Erlbaum Associates.

Sarason, S. B. (1996). *Revisiting "The culture of the school and the problem of change."* New York: Teachers College Press.

Schifter, D. (1997, April). *Developing operation sense as a foundation for algebra*. Paper presented at the annual meeting of the American Educational Research Association, Chicago, IL.

Spillane, J. P., & Thompson, C. L. (1997). Reconstructing conceptions of local capacity: The local educational agency's capacity for instructional reform. *Educational Evaluation and Policy Analysis, 19,* 185–203.

Tharp, R. G., & Gallimore, R. (1988). *Rousing minds to life*. New York: Cambridge University Press.

Warren, B., & Rosebery, A. S. (1995). Equity in the future tense: Redefining relationships among teachers, students, and science in language-minority classrooms. In W. Secada, E. Fennema, & L. Adajian (Eds.), *New directions for equity in mathematics education* (pp. 298–328). New York: Cambridge University Press.

Wenger, E. (1998). *Communities of practice: Learning, meaning, and identity*. Cambridge, England: Cambridge University Press.

Chapter 10

Teacher Collaboration: Focusing on Problems of Practice

David C. Webb
Thomas A. Romberg
Michael J. Ford
Jack Burrill
University of Wisconsin–Madison

In 1997, a group of middle school teachers in the Prairie Creek Middle School contacted researchers from the National Center for the Improvement of Student Learning and Achievement in Mathematics and Science (NCISLA) and expressed an interest in improving their mathematics and science program. This chapter focuses on the resulting successful collaboration.

In September 1997, we held an initial meeting with the interested mathematics and science teachers. After negotiations with administration, teachers and researchers collaborated on plans for the spring semester. In spring 1998, we began to work directly with four teachers (three at Grade 6 and one at Grade 8), with each teacher implementing at least one new unit from the algebra strand of one of the NSF-funded mathematics curriculum projects, *Mathematics in Context* (MiC; National Center for Research in Mathematical Sciences Education & Freudenthal Institute, 1996–1998). That summer three additional teachers (two more from Grade 6 and one from Grade 7) joined the project, and we held a 2-day summer institute in preparation for the school year. Some of the teachers also chose to meet with us at other times to develop new units of instruction for their classrooms.

During the 1998–1999 school year, teachers again implemented at least one new unit, either MiC or Boxer-based units. (Boxer is a flexible software system that supports student exploration of mathematics or science.) We intended to focus on ideas of modeling and student thinking, but in order to address issues relevant to teachers' concerns, we focused interchangeably on teachers' pedagogical concerns and instructional development of mathematical content. During the 1999–2000 school year, five Grade 7 teachers (persuaded by the sole Grade 7 teacher from the previous year) joined the project, and meetings focused on the teaching of MiC algebra units and the development of Boxer-based motion units. Problem-solving assessments were administered to all participating students in September, January, and May in the second and third year of the collaborative.

During this 3-year period, the pre-algebra curriculum in Grades 6 and 7 changed from a tracked program with a fragmented collection of activities to a more coherent and comprehensive mathematics program for all students. Teachers drew on the instructional expertise of colleagues and the research perspectives of the university team to explore ways to "uplevel the mathematics standards" in the middle grades. As teachers discovered new instructional methods that worked with their students, they shared these experiences with their peers, and, in many cases, instructional adaptations spread through conversations, both formal and informal, focused on students' mathematical thinking. As technical knowledge and discoveries were shared, the instructional quality of participating teachers was influenced in powerful ways.

In the sections that follow, we discuss the background of this successful project, examine issues in teacher practice, and close by talking briefly about the resulting impact on student achievement.

BACKGROUND TO THE COLLABORATIVE: DISTRICT LEADERSHIP AND EMERGENCE OF TEACHER COMMUNITY

District Leadership

District administrators were anxious to implement the district's instructional standards, which teachers had crafted in 1997, but the middle-grades mathematics teachers had major concerns: Sixth-grade teachers wanted help building on the background of the many students who had used innovative curricula in elementary grades, and all the teachers wanted to be able to provide both quality experiences for "all students" (which meant the school's tracking of students for mathematics needed to be reconsidered) and appropriate pre-algebra experiences in Grades 6 and 7 in preparation for algebra in Grade 8.

The collaborative was initiated with the full support of the district administration (including the district Director of Instruction and the Learning Resource Coordinator) and the middle school principal. Administrators saw the district in the process of changing their mathematics program from a traditional arithmetic program with ability grouping, toward an integrated program with heterogeneous grouping. They believed in a long-term collaborative approach to change, expected teachers to work toward agreed-on goals over time, and trusted teachers to make informed, reasonable decisions about ways to proceed. The Learning Resource Coordinator directly assisted the teachers and kept them informed about activities and decisions. Administrators also believed that the collaborative would help teachers broaden their scope about mathematics and mathematics instruction, and that, as a consequence, teacher-leaders would emerge. These new leaders working with their colleagues would, in turn, contribute to a change in the culture of what it meant to be a professional teacher in the Prairie Creek schools.

The innovative activities of teachers in the collaborative were also supported by district administration through their hiring practices and communication of goals to teachers not in the collaborative. Several teachers remarked (verbatim), "This district hires teachers that are overachievers and expects teachers to figure' out the rest." This shared perception led to a shared expectation that teachers would continue to experiment with, identify, and work to achieve best practices in their classrooms.

This district-based supportive approach to change in mathematics instruction needs to be seen in terms of issues that were facing the school district. First, the district population had recently changed from a stable mixed rural and bedroom community to a rapidly growing suburban community with considerable social diversity. (This population growth resulted in the building of a new middle school, which opened after this collaborative project ended.) The need for teacher-leaders was seen as important to the development of coherence. Second, needed bond proposals were typically passed only after repeated rejections, and in this case, a small but vocal group of parents was not pleased with the changing demographics and the needed resources. In response, some members of the community created a charter elementary school using a back-to-basics curriculum. Third, this district ranked among the best in the state and the top in the region in mathematics on the Grade 8 state assessment. They saw maintaining that high rank in the state as the challenge and noted the need to implement instructional standards aligned with the state and national standards. Fourth, the district mathematics committee established a goal to provide all students with ideas and skills central to algebra by the time they finished the eighth grade. In the final year of the project, this "algebra for all" initiative was later modified due to the adoption of a 3-year, integrated approach to mathematics for Grades 8–10, intended to provide the equivalent of the traditional algebra, geometry, and advanced algebra courses by the end of Grade 10. This integrated math program was to be implemented starting with Grade 8 in fall 2000. Our work with teachers on pre-algebra instruction proved to be a critical step in facilitating high school curriculum decisions. Finally, although the district was focused on the mathematics program, the science program was undergoing a similar shift.

The district administrators, although perhaps unfamiliar with all the details of the collaborative project, saw that "the culture [of the math department] has changed," as a district administrator commented in an interview (April 18, 2000). Teachers took responsibility for developing a coherent mathematics program *collaboratively*, and they recognized the strong connections of the work at Grades 6 and 7 with the work both at the elementary grades and with the new integrated math program at Grades 8, 9, and 10.

We also note that the programs in both mathematics and science prior to this project could best be described as "fragmented." There were, for example, different learning goals in place for different students (approximately 25% of the students were taught algebra in Grade 8). The instructional programs of each teacher varied in content covered, text materials used, methodologies for determining achievement, and so forth.

Growth of Professional Community

The coherent community of teachers we saw developing was built by the teachers themselves from the bottom up. With no appointed leader taking steps to direct the activities of individual teachers, the community emerged as a result of teachers' efforts to reform classroom instruction. Primary among their reform efforts was the articulation of elements of effective instruction, which occurred as teachers were introduced to the MiC and Boxer materials and again when they later used the materials with students. These instructional materials directed the activities of both students and teachers in ways that emphasized collaboration and discussion rather than rote memorization of facts and individual deskwork. Teachers found that in order to manage and direct discussions, they had to focus on student thinking while keeping the instructional end points in mind, a task they admittedly found particularly challenging.

The collaborative managed monthly meetings of researchers and teachers that explicitly addressed salient elements of instruction. As teachers shared experiences and began to articulate issues that arose in their classrooms, they were provided a vocabulary with which to discuss classroom practice. Several teachers remarked that the articulation of these elements legitimized and "professionalized" the intuitive ideas that had guided their practice prior to their participation in the collaborative.

Teachers also began meeting informally in smaller groups to discuss their experiences teaching specific units. They found that this informal collaboration provided an important source of support in that teachers could provide each other useful feedback about specific issues and content. Also, because many teachers felt the switch to a new form of instruction "risky," their collaboration and "sharing" served to reassure them that their decisions were good ones.

This practice-centered informal communication also supported authority and leadership at the department level. As noted earlier, the middle school mathematics program was the focus of school reform throughout the collaborative. A goal of this reform was to provide students with ideas and skills central to algebra by the time they finished the eighth grade. Teachers who had rarely expressed opinions in such a forum began taking more active roles in the decision-making process. One teacher in particular organized all other teachers in one grade to teach at least partly from MiC units because she believed so strongly that the structure and content of the curriculum met their needs. We believe that the emergence of formal and informal leadership stemmed from teachers' developing sophistication about effective mathematics instruction.

Classroom observation reports and teacher interviews suggest that growth in teacher practice occurred within the context of partnerships, in which teachers voluntarily worked with another teacher or a researcher to explore a pre-identified area of study (e.g., algebraic representations, the study of motion, classroom assessment). Through these partnerships, teachers grew more innovative and overcame significant obstacles in achieving desired classroom practices. These partnerships provided a safe context for experimentation, focused attention to classroom practice, peer and researcher observation, and shared reflection. The challenges teachers faced in designing classrooms that promoted understanding were often defined according to their conceptions of mathematics, conceptions of student learning, and pedagogical abilities. These partnerships gave teachers a safe context in which to challenge these conceptions and reconstruct their classroom practice.

Although a main feature of this research project was the teacher–researcher collaborative, finding that most teachers valued collaboration with their peers is not remarkable. Yet it is worth noting that teachers did not always take advantage of opportunities for collaboration, even when schools supported such activities. During the last year of the project, for example, school administrators encouraged teachers to observe each other, but teachers chose not to do so. Structural issues also limited the collaboration. At the end of our work with this district, teachers continued to meet in informal ways, but the physical separation that occurred when the middle school district moved about half of the teachers to a new school decreased overall collaboration. The middle school also reorganized teachers into teams or "houses," designed such that all teachers within a house taught different subjects to the same students, thus structuring teacher collaboration around students, not around instructional content. Clearly, such a focus has its advantages, but the emergence of informal "teams" of teachers during the collaborative suggests that instructional content-centered collaboration is also helpful for teachers, particularly during reform efforts.

TEACHER FOCUS ON PROBLEMS OF PRACTICE: EVIDENCE FROM CASE STUDIES

During our work with teachers, the growth in their knowledge of student reasoning and mathematical content supported a shift in their instructional approach and practices away from mere topic coverage toward facilitation of "sense-making" by students. In the second and third year, two recurring themes emerged: the overcoming of the "risks" entailed when engaging in student-centered pedagogy and teacher interest in discussing and exploring methods to assess student understanding. To elaborate on these *problems of practice*, we draw here on case studies of participating teachers. In general, teacher participation in these studies required a 4-to-8-week partnering with a researcher, who administered additional interviews, completed up to four classroom observations per week, and supported teacher reflection on classroom events and examples of student work.

Emerging Issue: Working Toward Student-Centered Pedagogy

For learning with understanding to occur on a widespread basis, students need opportunities to (a) develop appropriate relationships, (b) extend and apply their mathematical knowledge, (c) reflect about their own mathematical experiences, (d) articulate what they know, and (e) make mathematical knowledge their own (Carpenter & Lehrer, 1999). To teach for understanding, teachers need to design learning environments in which students have these opportunities to make sense of content and come to "own" what they know and learn (Bransford, Brown, & Cocking, 1999). But even when teachers recognize the need to change classroom practices, they often encounter significant difficulty when they attempt to orchestrate learning opportunities that are more student centered (Fennema & Nelson, 1997; Fennema & Romberg, 1999). As the following case summaries demonstrate, however, even though these difficulties can be quite similar, the ways that teachers resolve them can be distinctively different.

The Case of Beth Resnick.   Ms. Resnick had been a middle-grades teacher for almost 20 years. During the collaborative, she taught both mathematics and science. Her primary reason for participating in the collaborative was that she was dissatisfied with the science curriculum, which was organized by topic (e.g., "the solar system," "rocks and minerals," "weather"). According to Ms. Resnick, students did not have the opportunity to pose issues worth investigating or to make choices along the path of inquiry. The predetermined nature of the textbook (focused on sets of steps that students followed to generate conclusions from "experiments") removed any student motivation to engage in issues of interest or to develop deeper understanding of a textbook topic.

Participation in the collaborative led Ms. Resnick to shift from traditional, lecture-driven, direct instruction to pointed use of questioning, attention to student's ideas, and a demand that students articulate their own reasoning. In interviews, Ms. Resnick stated that this shift was not an easy one to make. Opening the lesson plan to the directions that students' ideas might take implied a certain amount of risk. She could not predict what students would say and often felt as though she had lost control of instruction. More generally, she found the new role as facilitator of discussions awkward, but found useful the sense of herself as a coach. With the new instructional style, however, Ms. Resnick found that students were learning more. By opening up discussions with the norm that all well-supported ideas were welcome, the variety of ideas presented increased greatly, students experienced more complexity, and what students articulated facilitated her identifying and helping students who were having trouble.

At the end of the collaborative, Ms. Resnick's goals were to develop better ways to encourage students to be aware of the reasons they believed what they did. She found it difficult to find ways to bring out in students a conscious awareness of their reasoning as well as to provide them with feedback that would direct them toward productive ways of thinking about a problem without giving them the answer. Ms. Resnick's work with the "physics of motion" unit illustrates some of the themes of this transition. In the interest of developing a unit that made clear the connection between motion and algebra, one of the authors of this chapter conducted weekly after-school Boxer enrichment workshops for interested sixth-grade students in the 1999–2000 school year. An adaptation of this sequence was then taught in the spring semester. During these sessions, students observed constantly accelerated motions, programmed these motions in Boxer, and drew representations of the motions with pencil and paper. Through their work, students faced important issues such as conceptualization of variables, connection of these variables to invented representations, quantification of these variables through measurement, and the separation of "signal" from "noise" in the resulting data.

Prior to participating in the development of the motion unit, Ms. Resnick had believed the best way to teach a physics concept such as acceleration was to introduce the concept directly and then to illustrate it with activities. In the motion unit, however, students were presented with motions with which they were already familiar: a ball dropped from a height of about seven feet; a ball rolling across a table and falling off to the floor; a shoved book that slid for a bit then slowed to a stop. Ms. Resnick demonstrated a particular motion for the students to observe and then asked them to draw the motion as well as they could with pencil and paper (see Fig. 10.1). The task to depict motion in a static medium required the invention of representational tools and was intended to focus students on ways of making their posited patterns explicit.

Initially, Ms. Resnick was afraid that the students would not see anything interesting in this task. After it was decided, for example, that the book slowed down, what else would there be to discuss? She was, therefore, surprised at the variety of ideas students had for representing the motion and at the symbol systems they developed. Presentation of students' representations led to further inquiry: Why is one symbol system better than another? Are all of the representations comparable to each other, or is there an element of incommensurability? What kinds of trade-offs are there from one representational technique to another? Is there a way to decide which is the best representation? Students ultimately had to grapple with the idea that what was considered "best" was relative to what they deemed important.

Ms. Resnick later traced her change in practice to the use of the reform curricula, which focused on student thinking, and the formal and informal collaboration with peers and researchers. She worked closely with the research team on the motion unit and also initiated a long-term collaborative relationship with Tracy Wilson, the other teacher who used the motion unit. This relationship developed to the point that they planned their math and science lessons together.

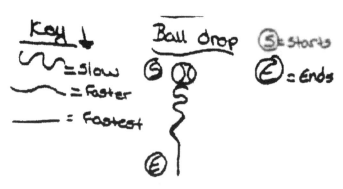

FIG. 10.1   Student representation of a dropped ball.

The Case of Jack Stoughton. Mr. Stoughton's most intense involvement with the collaborative was in the second year, when he and a researcher designed a 4-week Boxer-based geometry unit. This unit gave students the task of writing Boxer programs that would draw complex geometric shapes by iterating the rotation of simpler shapes. Mr. Stoughton stated that through the Boxer activities and subsequent class discussions, students learned relationships among important geometric concepts, such as interior angles, exterior angles, and regular polygons. In contrast to other geometry curricula, this unit engaged students, allowed them the freedom to develop their own tasks, and encouraged them to learn from each other as they worked in groups. As he shifted his focus to student thinking, the classroom environment also shifted, becoming more informal. Students had greater freedom to express their thoughts and approaches to problems, which in turn facilitated his assessment of their understanding.

Mr. Stoughton also thought it important that students not see math as something foreign to everyday life, but as something that had applications that could "empower" them in their day-to-day lives. He felt that listening to students was important because it made them feel valued as individuals. At the end of 2 years, he still felt he needed to strengthen his ability to elicit and understand his students' reasoning and to keep focused on the need to view each student as an individual with unique strengths and weaknesses. In the third year, the value he held for listening to students shifted to "helping students develop an awareness of how they and their peers are solving problems." He believed that excellent teaching resulted in students being willing to "take risks," and he made presentation of ideas to the group a class norm so that collaboration might lead to deeper thinking.

Over time, Mr. Stoughton recognized that the logistics of focusing on student thinking was difficult, particularly with a large number of students in the classroom. Although laudable, meeting each individual's instructional needs in a classroom of 25 students was, in practical terms, impossible. As a result, he shifted his goal from attempting to meet individual needs to identifying common problems, but acknowledged that defining such commonalities was a challenge. Similar to most teachers in the collaborative, he found the sharing of instructional techniques with other teachers essential and identified the lack of time for peer collaboration as the main challenge to realizing excellence in teaching.

Summary. As the cases of Ms. Resnick and Mr. Stoughton show, some teachers gave considerable effort to designing student-centered learning environments. Yet, as teachers reexamined what it meant to learn and understand mathematical ideas, they struggled to create learning environments consistent with their new assumptions. Practical considerations often resulted in trade-offs between meeting the needs of individual students and the practical constraints of teaching a classroom of students. To some degree, teachers were able to resolve the difficulties they encountered, although all agreed much work remained ahead. For each teacher, the collaborative positioned experimentation with practice as a worthy endeavor and promoted individual persistence in moving beyond initial uncertainty and discomfort. The collaborative also provided some degree of accountability, in that the move toward student-centered pedagogy was a collective endeavor.

Emerging Issue: Developing Principled Classroom Assessment Practices

Through the first two years of the collaborative, teachers' excitement with the new curricula (i.e., MiC and Boxer) was tempered by a growing sense that their methods for assessing student learning were inappropriate to the materials. In response to issues raised by teachers during monthly meetings in the second year and in the end-of-year teacher interviews, the research team decided to organize the meetings in the third year around two themes: (a) the identification of "core concepts" for assessing the development of algebraic understanding across all grades and (b) the discussion of alternative methods for designing, scoring, and grading formal assessments. To illustrate contrasting approaches resolving issues in classroom assessment, we focus on two teachers' efforts to develop a more meaningful approach to assessment.

The Case of Judy Koster. Ms. Koster had taught K–8 students for over 20 years. During the collaborative, she taught seventh-grade mathematics, language arts, and social studies. After hearing sixth-grade teachers talk about their experiences using MiC, she decided to participate in the collaborative. Dissatisfied with the dryness and repetition of teaching with a textbook, Ms. Koster wanted to experiment with a curriculum that challenged students *to reason* about math rather than simply to memorize computation methods. As she grew more familiar with MiC, she also realized that the way in which MiC spiraled algebraic concepts from unit to unit provided a developmental basis for instruction sequencing that she had tried to create in prior years, but had been unable to construct on her own.

To Ms. Koster, the greatest challenge was the design, selection, and scoring of performance assessment tasks, but her persistence in addressing assessment issues in the monthly meetings also reflected the concerns of her colleagues. Even though she had developed her own method of rubric-based scoring of student work, she found the scoring of more challenging assessment tasks problematic and experimented with various ways to implement assessment that was fair, informative, and practical for both students and teachers. In contrast to the ubiquitous protocol for testing, in which students are not allowed to ask questions about the test problems, Ms. Koster decided to engage students in a brief reading and discussion of the problem context in order to mediate limitations in students' reading ability or lack of experience with a particular context. As evidenced in classroom observations and interviews during the third year, Ms. Koster was also less concerned with the *number* of problems covered or assessed than with the ways students articulated their understanding of mathematics. As Ms. Koster explained:

> It goes back to that whole explanation of, do you need 20 problems from a kid to understand that they understand it? Or can you put your energy and your focus in your discussion into one word problem? I think it's a tremendous shift in thinking and trade-off, but my, my main suggestion would be, put emphasis, focus, and energy on one or two problems. The exhausting part of this is you're on deck the whole time. But the trade-off is, if you're tuned in, I think you will have a more immediate sense of how the kids are doing. (Interview, May 6, 1999)

Ms. Koster noted that her participation in the collaborative made her excited about being "a student again"—solving problems and discussing with researchers and colleagues ways to use activities in the classroom. The discussions in the monthly meetings affirmed her philosophy of how mathematics should be taught and gave her the opportunity to relate her views to contemporary research in mathematics education. The collaborative also reinforced the value of setting aside time to collaborate with other teachers.

Ms. Koster and Deborah Harley, her "house" partner, collaborated in planning lessons, designing assessments, and scoring student work. During the 1999–2000 school year, they used the same formal assessments, administered tests the same way, and used a similar rubric to score student responses to assignments and tests. When using the end-of-unit assessments, both Ms. Koster and Ms. Harley decided to use the tasks as additional learning opportunities instead of as time-restricted tests of student knowledge. On the first day of a unit test, the entire class discussed ways to solve each problem, much as had been done throughout the unit. On the second day, students received a blank copy of the same assessment and were asked to write individual responses to each problem without the benefit of classroom discussion. In this way, Ms. Koster and Ms. Harley were able to informally assess student thinking during the pre-discussion and identify student misconceptions still present in students' work on "test day." We note that, although their approach gave the appearance of coaching students during formal assessment, it did not compromise student performance on the third-year algebra assessment we administered to all sixth- and seventh-grade mathematics classes (Webb et al., 2001).

The Case of Rebecca Mauston.   Ms. Mauston had taught middle-grades mathematics, social studies, and language arts for 6 years. During the collaborative, she taught sixth-grade mathematics and social studies. When Ms. Mauston initially attended monthly meetings, she found that she enjoyed the opportunity to discuss the issue of practice with her colleagues, and she became more intrigued with MiC. She realized that the materials and activities of the collaborative paralleled much of what she was interested in accomplishing in her own classroom: creating a classroom community in which students respected each other and felt safe enough to share their thinking. In the final year of the collaborative, Ms. Mauston began to experiment with new methods of assessing, documenting, and grading growth in student knowledge over time. During several meetings with one of the authors of this chapter, Ms. Mauston outlined a more principled methodology for assessing student thinking.

Central to Ms. Mauston's change in practice was her development of a multilevel framework for assessing student reasoning in each unit of instruction. During her preinstructional planning, she established assessment goals for the unit, identified benchmark tasks (from tasks included in the unit and other available resources), and decided the appropriate social context for using these tasks. Her instructional goals were a combination of district content standards, learning objectives outlined in each unit, and personal interests. Ms. Mauston selected benchmark tasks on the basis of the accessibility of the task to a range of informal and formal student reasoning, the relevance of problem context, and her judgment of which tasks best represented her instructional goals. She used assessment tasks in a range of social contexts to create opportunities for dynamic assessment (Brown & Ferrara, 1999; Campione & Brown, 1987), in which student performance could be assessed with various degrees of instructional support, such as during whole-class discussion, group work, semistructured interviews, or during individual seatwork. As part of her new method of assessment planning, Ms. Mauston instituted student participation in "math chats" in order to take advantage of assessment opportunities available during students' mandatory tutorial period.

In assigning grades, instead of computing a percent-based grade, she rearticulated what students should demonstrate to earn a particular grade. Ms. Mauston organized learning objectives according to three levels of student understanding: basic skills, application, and analysis and extension. She then designed a rubric that assigned grades according to student performance on tasks at these levels. To earn a "C," for example, students needed to demonstrate most basic skills, good homework habits, and moderate participation. To earn a "B," students needed to demonstrate all basic skills, earn partial credit on most application problems, and maintain good homework habits and class participation. To earn an "A," students needed to extend their mathematical knowledge and demonstrate higher levels of reasoning on a higher level task.

In this process, Ms. Mauston made the meaning of grades and the ways they were earned explicit to herself and to her students. Classroom observations and interview data suggest that after Ms. Mauston introduced these assessment practices, her instruction and classroom discussions became more focused, and student motivation for learning mathematics increased. In a series of follow-up student interviews in the final year of the collaborative, Ms. Mauston's students described quizzes, tests, and "math chats" as opportunities to demonstrate their understanding rather than as a way to "get a good grade." By communicating curricular goals to students and designing a system that could be understood by students and flexible to their needs, Ms. Mauston encouraged students to assume greater responsibility for their own learning. As one sixth-grade student noted, "[In the new system] you actually have to show that you know the skill. It is not like in a multiple choice test, and you get it right, luckily, or something like that. You have to prove it to her that you know it and demonstrate it."

Ms. Mauston's assessment program caught the attention of the Director of Instruction, who asked her to participate in district-level meetings to write guidelines for application of district curriculum standards. In summer 2000, sixth-grade mathematics teachers in the Prairie Creek School District reviewed her methods for assessment planning and grading and agreed to develop similar assessment frameworks for use in their classrooms. District administrators encouraged these meetings and supported the planning needs of the sixth-grade mathematics teachers through paid time for summer and after-school monthly meetings. During the 2000–2001 school year, similar efforts at improving classroom assessment were promoted by the district through several staff development meetings to disseminate assessment guidelines for rubric-based scoring and grading and to allow other teachers to share other innovative assessment practices. In a relatively short period of time, the district attempted to institutionalize one teacher's innovation as a cross-disciplinary model for assessment.

Summary.   We note that not all teachers in the collaborative adopted such assessment practices. Some colleagues questioned the validity of these methods and were uncomfortable with ways students could abuse assessment methods that included opportunities for students to collaborate with peers, share solution methods, and revise their work. On the other hand, teachers that continued to use conventional assessment practices were not able to resolve the disparity between their personal assessment of what students knew and the results they obtained from their more conventional routines for assessing, scoring, and grading student work. As Mr. Vandenberg, a seven-grade teacher, noted:

> I knew in my heart and my gut who was getting it and who wasn't. But in terms of, "Okay, how do I put that into a grade?"—that was a whole other challenge. We talked about it, and you have to give yourself permission probably [to] not grade as much. I mean things that you do grade, you have to get a rubric, or you have to get something down on paper so the kids really understand how they get the grade. (Interview, June 20, 2000)

IMPACT ON STUDENT ACHIEVEMENT

Student achievement data were gathered on all students taught by participating teachers during the second and third years of the collaborative. In the second year, 1998–1999, three different assessments for each of the three grade levels (specifically developed for this study) were administered at the beginning, middle, and end of the school year. In the first of three problem contexts, students were asked to continue a given pattern and provide an example of a similar one. The second problem context involved a skateboard race and addressed aspects of motion such as relative position, velocity, and acceleration. The third context was designed to elicit students' informal solutions to covariation problems that involved different combinations of menu orders, rental fees, and campground areas. Each context included questions to elicit student explanations of their solution strategies.

Many of the tasks used in the physics of motion and algebraic reasoning contexts were quite challenging and required students to interpret and reason with representations rarely addressed in the middle grades. Importantly, we found that students who engaged in the problem contexts were able to do so with a relatively high degree of success. The assessment results also suggested the following patterns of student achievement for the second year of the collaborative:

- Most incoming sixth-grade students were able to complete a given pattern.
- Seventh and eighth-grade students showed progressive growth in informal and preformal algebraic reasoning and were able to generalize the pattern described in a realistic problem context.
- Comparison of teachers' curricular decisions and student achievement data showed that student opportunity to learn significantly af-

fected student performance in preformal algebraic reasoning and generalization.

- The cooperation and mutual support of the sixth-grade teachers led to a more coherent mathematics program and increased opportunities for students to engage in and improve their solving of complex, nonroutine problems. By the spring, sixth-grade students were beginning to perform at a level equal to the performance of eighth-grade students.

In the third year, 1999–2000, the teachers emphasized the study of student reasoning in algebra and sought to provide students with appropriate pre-algebra experiences in sixth and seventh grade. In alignment with this goal, the third-year assessment was designed to document student achievement in various aspects of algebra, including use of patterns, covariation, and understanding of exponential phenomena. The first section of the assessment included five multiple-choice questions selected from eighth-grade public-release items used in the Third International Mathematics and Science Study (TIMSS). The second section of the assessment included four contexts with several constructed-response prompts associated with each problem context. Three of these problem contexts were used on the previously described second-year assessments. Another problem context was taken from the Grade 7 Problem Solving Assessment (Dekker et al., 1997–1998) developed for the "Longitudinal/Cross-Sectional Study of the Impact of Mathematics in Context on Student Mathematical Performance." Each context included prompts to elicit student explanations of their solution strategies. (For a more detailed report of data from this assessment, see Webb et al., 2001.)

Overall, the student achievement results for the third year demonstrated substantial growth in students' algebraic reasoning. We attribute these results to teachers' emphasis of algebra content in sixth and seventh grades. More specifically, the assessment results showed the following patterns of student achievement:

- In the longitudinal analysis, we found that the differences in student performance from year to year were statistically significant. In two cases, the difference in student performance was greater than one standard deviation. However, there remains significant room for improvement on many of these items.
- On the multiple-choice tasks, we found an overall difference in performance between sixth- and seventh-grade students. Similar to achievement patterns found in the second-year data, we conjectured that student opportunity to learn and instructional emphasis significantly affected student performance on these items.

- In contrast to this difference, we found student performance on particular items in Grades 6 and 7 to be equivalent. In order to explain why performance on these items was equivalent when all other items saw performance differences across grades, we also cite student opportunity to learn.
- In Grade 6, we found a significant difference in performance between those students who used preformal algebraic tools (e.g., combination chart) and those who did not. Use of such tools, however, was not found in responses of the seventh-grade students who successfully responded to this item. Very few seventh graders organized their problem solving with such tools and instead demonstrated use of formal, symbolic approaches.
- We found variation in student performance across teachers that taught the same MiC units. Although some of these differences could be attributed to a general teacher effect, based on observations and informal conversations we had with teachers, we conjectured that teachers' differential instructional emphasis within the same units had an impact on student performance. For instance, some seventh-grade teachers explored exponential growth and decay phenomena in detail, whereas others skimmed the topic.

Students' opportunities to learn and teachers' instructional emphasis had a significant impact on student achievement. However, simply stating that "students learned what teachers decided to teach" misrepresents the relationship between teacher change, curriculum coherence, and student achievement that we observed in this study. Providing students opportunities to learn mathematics with understanding required an environment in which teachers had an opportunity to develop pedagogical knowledge, mathematical knowledge, and classroom practice. By the end of the third year, five of the six Grade 7 teachers were in their first year of using a new mathematics curriculum. Every teacher in the collaborative noted that his or her prior classroom practices were challenged in one form or another. We view the positive trend in student achievement data to be a set of assessment snapshots that confirm the improvement in student learning of mathematics that occurred during the course of the collaborative. The longitudinal growth in student performance in the third year is particularly significant given that most of the teachers for this student cohort were just beginning to grapple with the inherent challenges of student-centered instruction and more ambitious forms of classroom assessment. Over time, as teachers continue to improve their classroom practices and become more familiar with the way mathematics concepts are developed, we expect their students will demonstrate even greater performance on similar assessments.

CONCLUSION

The collaborative created a "safe" environment for teachers to share and discuss their problems of practice. As they made their practice public, they found that some issues, which they might have perceived as idiosyncratic, were often shared by many of their colleagues. Through the collaborative, "personal" teaching issues were often reflected back as *problems of practice*, and teachers, supported by the district and working jointly with researchers were able to experiment with changing their practices through reform approaches to instruction and assessment. Certainly, the growth of teachers as they participated in the collaborative varied considerably. Some, like Ms. Resnick and Ms. Mauston, experienced considerable change in their work as professional teachers. Some, like Ms. Koster, found that the instructional expectations reinforced their beliefs about excellent teaching. All found the teaching of reform units to be challenging, but all also saw the beneficial impact of such instruction on student learning. In the sections that follow, we summarize the effect the collaborative had on their instruction and on their implementation and use of assessment.

Collaborating to "Uplevel" Instruction

As teachers piloted the new units (MiC and Boxer-based), they deepened their understanding of student conceptions of math and science. As teachers continued to elicit student explanations of their strategies, sometimes taking the stance of someone completely ignorant of a mathematical topic to encourage student explanations, they found that their own understanding of the concepts they were teaching was enhanced by students' responses. As a result, teachers began to make more-informed decisions about which sections of a unit they should emphasize, skim over, or avoid. As they became more familiar with the content of the units, their instructional stance became more proactive, and they began to interject classroom discourse with more formal mathematical language in order to connect student conceptions to a shared language.

Teachers' experimental use of Boxer and MiC, for example, permitted a "safe zone" for teachers to explore content not previously used with students at these grade levels. As teachers piloted Boxer and MiC algebra units, they were intrigued by the enriched nature of students' mathematical communication and reasoning. Because teachers often experimented with the same instructional units, they were able to discuss not only the interaction of curriculum, instruction, and assessment in a shared context but also the merits of a particular activity and the range of student responses their instructional practices had elicited.

Teachers were quick to identify this peer collaboration as key to supporting their efforts to implement reform practices and in helping them resolve the practical problems that arose as they worked to reform their practice and to develop the skills they felt they needed. Sharing strategies and conceptions of the desired skills tended to "spread out the risk" the teachers were taking in implementing change. The reforms supported by the collaborative necessitated a modification of the traditional model of the classroom, in which "correct" information is transmitted from an expert to a roomful of novices, toward a more open-ended, student-centered model in which knowledge is dynamic, collaboration is involved in cognition, and students and teachers sometimes share leadership roles. The difficulty faced by most participants in implementing this model seemed rooted in two challenging issues:

- Planning out an instructional route that would enable them to simultaneously focus on student discussion, infer student thinking, and gather evidence on student understanding.
- Shifting from the comfortable role of class leader to the "risky" role of class facilitator.

Effecting these changes was difficult, but as teachers shifted to the focus on student thinking inherent in this reform model and experimented with new ways to scaffold student learning, many realized that this was what they had been trying to do for years. Teachers already knew from experience that students varied considerably in their reasoning about problems, but what students thought and the strategies they used were hidden from traditional approaches to instruction and assessment. Reinforced in their efforts by the district, the researchers, and their peers, teachers experimented with practices in actual classroom environments. Discussion of what they did and observed at the monthly meetings legitimized their intuitions and fostered their sense of professionalism.

Linking Assessment and Achievement

Three concurrent developments emerged in teachers' classroom assessment practices: the investigation of instructionally embedded assessment; the selection, design and use of assessment tasks; and the scoring and grading of student work. Both formally and informally, teachers in the collaborative began to discuss their assessment, scoring, and grading approaches and shared their experiences with assessment tasks in MiC and the challenges they faced in using tasks to assess student reasoning and problem solving. At their monthly meetings, teachers deliberated the pros and cons of different approaches and talked about assessment methods used in language arts, science, and social studies. The framework for planning classroom assessment developed by Rebecca Mauston was shared with teachers outside of the mathematics department and was adapted for use by language arts and social studies teachers. During the final month of the collaborative, Grade 6 mathematics teachers received district funding to develop MiC unit assessment plans, a project that included the identification of instructional goals in light of the district's standards and the selection of unit benchmark tasks. Several teachers found that their use of reform assessment practices and the identification of appropriate informal classroom assessment opportunities had influenced their instructional decision making, the selection of curricula, the selection and modification of assessment tasks (and how they were used), and even the instructional pathways they used to guide student discourse.

Certainly teachers found that their change in practice deepened and changed their instruction, but the change in their classroom assessment practices also had a notable impact on student disposition toward learning mathematics. Instead of using assessment primarily to judge whether students had successfully *completed an activity*, teachers had begun to use assessment to evaluate whether students *understood mathematics and science concepts*. When one teacher adjusted her expectations of the ways students should demonstrate their understanding of mathematical skills and concepts, her students improved their work according to the expectations they had *for themselves*. For example, when students were presented with a range of mathematical competencies they needed to demonstrate to earn an "B," students took greater ownership in the learning process and decided which concepts they wanted to learn and demonstrate on the targeted assessments.

As teachers created contexts to support student learning of mathematics and science with understanding, they discovered that middle-grades students could model fairly sophisticated phenomena. Sixth-grade students were able to represent cases of constantly accelerated motion in a variety of ways, and these representations supported discussions about concepts central to kinematics (e.g., speed, time, distance). Overall, students were also able to arrive at multiple conceptual organizations of concepts such as accelerated motion on a ramp (in order to measure changing speed) and could recognize representation issues and could refine their depictions accordingly. Importantly, students were able to demonstrate that they had learned to support what they found through argumentation.

Summary

Overall, the collaborative was effective in supporting and sustaining the effort toward improving instruction and in building school capacity to provide all students with a mathematics program that was comprehensive and meaningful. The context of university collaboration also validated teachers' efforts to administrators and parents, provided incidental technical support, and supported teachers' efforts through provision of necessary material resources such as MiC units, Boxer software, and computers (which would not have been purchased by the district because of limitations in budget allocations for mathematics resources).

After the researchers pulled back, the project was sustained by further exploration of practice by project teachers, dissemination of support structures and classroom practices across the Prairie Creek School District, and "travel" of project findings to researchers and teachers in new projects and new school districts. Field notes gathered by researchers who continued to observe monthly meetings after the conclusion of the collaborative confirmed sustained deliberation of curricular decisions and instructional approaches in a continual effort to improve student learning. The level of teacher discourse remained focused on problems of practice, further suggesting that the norms of professional discourse developed during the collaborative were maintained, at least in the short term.

Clearly, to sustain growth in teachers' classroom practice, teacher collaboration in content areas must continue as a means to support further experimentation and development of teacher knowledge central to teaching and learning mathematics with understanding. Such collaboration could focus on teachers' (a) engagement in new activities that promote teacher learning of mathematics content, (b) recognition and collection of evidence of students' understanding and learning, (c) elicitation and interpretation of that evidence, and (d) exploration of assessment methods appropriate for assessing student understanding. Although such collaboration might include university personnel, over time we expect teachers to take greater ownership in maintaining collaborative structures and continuing and sustaining the discussion of teaching and learning mathematics.

We note in closing that the key issues—district leadership support and the emergence of teacher community—had great effect on the collaborative, which continues informally despite major changes in the district and schools themselves. We also wish to point out the importance of the shift in assessment practice, which—when used not to rank students, but to inform instruction—led to better quality of instruction and substantial increase in student achievement, as measured both by informal classroom assessment and external formal testing.

AUTHOR NOTE

The research reported herein was supported in part by a grant from the U.S. Department of Education, Office of Educational Research and Improvement, to the National Center for the Improvement of Student Learning and Achievement in Mathematics and Science (Grant No. R305A60007-98), and by the Wisconsin Center for Education Research, School of Education, University of Wisconsin–Madison. Any opinions, findings, or conclusions are those of the authors and do not necessarily reflect the position, policy, or endorsement of the supporting agencies.

REFERENCES

Bransford, J. D., Brown, A. L., & Cocking, R. R. (Eds.). (1999). *How people learn: Brain, mind, experience and school*. Washington, DC: National Academy Press.

Brown, A. L., & Ferrara, R. A. (1999). Diagnosing zones of proximal development. In P. Lloyd & C. Fernyhough (Eds.), *Lev Vygotsky: Critical assessments: Vol. 3. The zone of proximal development* (pp. 225–256). New York: Routledge.

Campione, J. C., & Brown, A. L. (1987). Linking dynamic assessment with school achievement. In C. S. Lidz (Ed.), *Dynamic assessment: An interactional approach to evaluating learning potential* (pp. 82–115). New York: Guilford Press.

Carpenter, T. P., & Lehrer, R. (1999). Teaching and learning mathematics with understanding. In E. Fennema & T. A. Romberg (Eds.), *Classrooms that promote mathematical understanding* (pp. 19–32). Mahwah, NJ: Lawrence Erlbaum Associates.

Dekker, T., Querelle, N., van Reeuwijk, M., Wijers, M., Feijs, E., de Lange, J., Shafer, M. C., Davis, J., Wagner, L., & Webb, D. C. (1997–1998). *Problem-solving assessment system*. Madison, WI: Wisconsin Center for Education Research, University of Wisconsin–Madison.

Fennema, E., & Nelson, B. S. (1997). *Mathematics teachers in transition*. Mahwah, NJ: Lawrence Erlbaum Associates.

Fennema, E., & Romberg, T. A. (Eds.). (1999). *Classrooms that promote mathematical understanding*. Mahwah, NJ: Lawrence Erlbaum Associates.

National Center for Research in Mathematical Sciences Education & Freudenthal Institute. (1996–1998). *Mathematics in context*. Chicago: Encyclopædia Britannica.

Webb, D. C., Ford, M. J., Burrill, J., Romberg, T. A., Reif, J., & Kwako, J. (2001). *Prairie Creek Middle School design collaborative: Year 3 student achievement data technical report*. Madison, WI: Wisconsin Center for Education Research, University of Wisconsin–Madison.

Chapter 11

Managing Uncertainty and Creating Technical Knowledge

Walter G. Secada
University of Miami, Florida

Tona Williams
University of Wisconsin–Madison

There is a very strong folklore among mathematics and science educators about the difficulties that teachers experience as they shift from conventional forms of teaching to teaching for understanding (see, e.g., Ball & Rundquist, 1993). Using data that we collected in a larger study of six research sites (Gamoran et al., 2003), we developed three interrelated hypotheses about the role of uncertainty in this process: (a) Uncertainty becomes salient within all facets of teachers' practice as they adopt teaching for understanding. (b) Teachers' professional communities are valuable sites for the management of that uncertainty. (c) New forms of technical knowledge are created as a product of those communities.

THE DILEMMA OF UNCERTAINTY

The shift toward teaching for understanding makes more explicit the uncertainty inherent in teaching in that the routines that help teachers make curricular decisions as they instruct and assess students (Fennema & Romberg, 1999) are lost or substantially altered. Even in its more traditional forms, teaching involves contingency in that instructional decisions are based on how well students learn content and how well the classroom is managed. Teaching is complex, risky, and ambiguous: It entails attention to many students simultaneously and to the classroom's overall ambience; it is not always successful, nor is it a given that all students will learn the content; and its goals are often multiple, and ways of judging success often contradictory.

In describing teaching as an "impossible profession," Cohen (1988a, 1988b[1]) positioned teaching as one of many human-improvement professions in which outcomes are problematic. In such professions, teaching in particular, the results and processes are neither clear nor unambiguously connected. The human-development professional depends on the cooperation of someone else for success, but obtaining the full participation of the nonprofessional partner—in this case, the student—is often difficult. In other words, teaching is uncertain because of its problematic outcomes and processes, its dependency on others (the students) for success, and the challenges that need to be faced in obtaining the full participation of those nonprofessional partners.

Uncertainty, however, is not a problem to be solved, but a dilemma to be managed. We note that not everything about teaching is problematic and uncertain. Expert teachers have a wide repertoire of routines that they apply in many different situations with a high probability of success. They know what works and why it works, and can implement backup plans when things go wrong. But teachers, expert and nonexpert alike, depend on one another to manage this uncertainty collectively. Colleagues within a professional group, for example, establish norms governing how new members join the group and share information, and what practices are acceptable. Within a school setting, individual teachers receive support in adopting practices that their departmental colleagues, a group of teachers within the larger school or others within the profession, have tacitly sanctioned if not actively pursued.

Mathematics and science teachers can reduce uncertainty through a variety of practices and strategies. Some of these practices might be more productive than others in terms of teaching for understanding. First, a school, department, or group of teachers can focus their instructional goals, more clearly specify the methods by which they will meet those goals, or both (Cohen, 1988a, 1988b). For example, a mathematics or science department's teachers might maintain a sharp instructional focus on the memorization of facts (goals) through drill and practice (the method for achieving that goal); alternatively, they might abandon any kind of basic-skills development (goals) by adopting a purely problem-solving approach in teaching

---

[1]Although not published in an easily available book or journal, Cohen (1988b) contains a version of David Cohen's observations about the uncertain nature of "adventuresome" teaching that updated his (1988a) piece, which had received feedback from colleagues in the study of teaching. Hence, this chapter draws on that version as opposed to the more readily available Cohen (1988a).

(method). Taken to an extreme, either of these goal–mechanism pairs would reduce uncertainty within a group of teachers, but they would not be productive in terms of achieving a balanced set of instructional goals that stressed student understanding.

Second, teachers could manage the uncertainty found in their practices by strengthening the match between their goals and tools on the one hand and their clients' demands and needs on the other (Cohen, 1988a, 1988b). For example, private schools and tutoring services sell a set of curricular goals and instructional methods on which the people who pay for those services agree. This selection helps to ensure that students, parents, and teachers will work together to achieve a common set of goals. What is more, if enough people refuse to pay for the service, the private school and its teachers face the choice of either modifying their goals or going out of business.

Less productively in our opinion, teachers in a school or in a secondary mathematics or science department could also increase the match between teacher goals and student demands by referring students who refuse to cooperate (or who are otherwise deemed to be failures) to lower tracks, low-ability groups, and special education services. More productively, teachers could exhort students to study, come to class, and in other ways comply with the school's academic mission. In a study of restructured schools, Marks, Secada, and Doane (1996) documented how students were able to understand and could repeat, in their own words, their teachers' efforts to focus the students on academic work.

A third way of reducing uncertainty would be to address how notions of success are defined (Cohen 1988a, 1988b). For example, teachers in a department could loosen how they think of the connection between teaching and learning. Teaching could become "what teachers do" in a manner that becomes increasingly disconnected from student learning; a lesson or unit would be successful if it were taught according to those specifications. In this scenario, teaching would become more and more of a certain practice. Meanwhile, learning would become the students' responsibility and students would be successful insofar as they learned the content. Taken to an extreme, teaching and learning could be thoroughly divorced from one another as in, "I taught this—it's your job to learn it."

Cohen does not mention a fourth method by which the uncertainty of teaching could be managed explicitly, although it is tacit in his observations involving individual teaching expertise. This fourth method would involve the learning and routinization of a range of teaching practices that are organized into coherent schemes through the development of teacher expertise. In other professions, such a body of knowledge and practices would be considered that profession's core. That core is typically referred to as technical knowledge, in contrast with the nontechnical knowledge that nonspecialists possess. At a minimum, the technical knowledge found in teaching would include content knowledge and pedagogical content knowledge, as described by Shulman (1987). A department that adopted strong norms of professional learning and development would support individual teachers' acquisition of such a body of knowledge and its associated practices.

TEACHING FOR UNDERSTANDING

Teaching for understanding in science and in mathematics is characterized by a focus on student thinking, attention to powerful scientific and mathematical ideas, and the development of equitable classroom learning communities (Gamoran, Secada, & Marrett, 2000). Many of the teachers whom we observed and interviewed in our larger study (Gamoran et al., 2003) reported experiencing difficulties and feeling uncertain and anxious in trying to decide what and how to teach for understanding. Teaching for understanding makes more salient the uncertainties that are inherently a part of teaching. When one veteran teacher in a Wisconsin suburban elementary school began to tie her decision making and the resulting practices more closely to student understanding, the tried-and-true ways by which she had been able to differentiate a good from a bad lesson were rendered problematic, which served to increase her sense of unease. She described the tension that she experienced as a result:

> This is my 15th year, and when I started, everything was pretty rote ... and teachers basically used the textbooks to guide their teaching.... [It] made me feel like I was in control of the classroom because I knew what we were going to do every day ... striving for teaching that is more meaningful for the students means giving up that control, and it means listening to what the students say and looking at what they can do and what they have trouble with ... So giving up the control is kind of hard for me. I think I am getting better at it because I have been practicing it. (Teacher interview)

A shift from conventional teaching practices to teaching for understanding makes uncertainty more prominent because it takes away from teachers (even expert teachers) the practices and routines that they have developed over years and that worked well according to an old set of criteria. Teachers in this situation sometimes wonder which practices remain viable and which ones no longer work. Many of the strategies that teachers could employ to manage this uncertainty are themselves under question.

In addition, as this teacher noted, there is no set of readily available routines based on teaching-for-understanding practices that are comparable to the routines of conventional teaching that teachers find they are abandoning. This occurs, in part, because teaching for understanding does not have the well-developed body of technical knowledge that is available for conventional teaching. Furthermore, although some very gifted teachers have developed practices under unique conditions or with strong support (e.g., Ball & Rundquist, 1993; Fennema & Romberg, 1999; Hiebert et al., 1997; Lampert, 1986, 1990), it is not clear whether these practices can be adopted (or adapted) by other teachers who are shifting their practice toward teaching for understanding.

Unlike conventional instruction, teaching for understanding requires teacher knowledge that goes beyond knowing the content of mathematics and science and the pedagogical content knowledge involved in teaching mathematics and science (Shulman, 1987). Teachers who teach for student understanding must also know how students *reason* mathematically and scientifically (Fennema & Franke, 1992). The lack of ready-made widespread routines that teachers can draw on imposes the added burden of creating and validating a new set of practices that support teaching for understanding.

USING PROFESSIONAL DEVELOPMENT TO MANAGE UNCERTAINTY IN THE CLASSROOM

How, then, do mathematics and science teachers perceive and manage the increased salience of uncertainty that comes with a shift to teaching for understanding? Organization theory (Rowan, 1990; Rowan, Raudenbush, & Cheong, 1993) suggests that as teaching becomes more nonroutine, teachers will increase their reliance on a network structure, in this case, their professional community:

> Nonroutine technologies require workers to engage in frequent searches for solutions to complex technical problems ... and as workers require more technical information to solve these problems, hierarchical and standardized approaches to work become inefficient. As a result, organizations develop lateral patterns of communication. *Network structures replace hierarchical structures of management, and technical work comes to be guided by information and advice received by colleagues rather than by centralized and standardized task instructions* [italics added]. In this situation, a system of ad hoc centers of authority and communication emerges, with those possessing relevant information and expertise assuming leadership no matter what their formal position of authority. (Rowan, 1990, p. 357)

We hypothesize that teachers' professional communities constitute Rowan's "network structures," which help them to manage the uncertainty that becomes salient as they shift from conventional forms of teaching to teaching for understanding.

In what follows, we explore three interlinked hypotheses:

1. The shift from conventional teaching practices to teaching for understanding makes teachers' uncertainty more salient in all areas of teaching: curriculum, instruction, assessment, and teacher knowledge about student reasoning.

2. Professional communities of teachers provide social mechanisms—in particular, through the process of what teachers call "sharing"—through which uncertainty can be managed. Sharing allows teachers to respond to one another's private affect (e.g., frustration), beliefs, and ideas; provide support and encouragement to try out new ideas in classrooms; and help each other to maintain practices that resonate with newly developing ideas about how to teach for understanding.

3. The practices related to teaching for understanding that teachers validate among themselves within their professional communities, the beliefs that undergird those practices, and the warrants for that validation together represent a body of technical knowledge that the professional community creates as an artifact of its efforts to manage uncertainty.

In exploring these hypotheses, we draw on collaborative efforts of a larger research team (Gamoran et al., 2003), which collected data from six research sites. At each of these sites, mathematics and science education researchers collaborated with teachers to foster teaching for understanding in classrooms. Between 1996 and 2000, for two to three years each, we monitored Wisconsin suburban elementary, middle, and high school sites; a Massachusetts urban elementary school site; a Tennessee urban middle school site; and a Wisconsin urban high school site. We utilized 155 interviews with teachers who participated in the design collaboratives, 42 interviews with district and school administrators, and 102 observations of design collaborative meetings. Incorporating both pre-specified and emergent coding categories, we conducted an interpretive analysis, following a multisite case study approach.

Our investigation of the role of uncertainty in these collaboratives begins with examples drawn from across the six sites. Then, we utilize a case study of the Oberon High School science design collaborative to explore in greater depth how participants recognized and managed uncertainty.

ROLE OF UNCERTAINTY: EXAMPLES FROM THE SITES

Increase in Salience of Uncertainty

Across the six research sites that were part of Gamoran et al.'s (2003) larger study, we found ample evidence that, with the shift toward teaching for understanding, teachers more acutely experienced uncertainty involving curricular decisions, their instructional practices, classroom assessment of students, and ways of figuring out how students reason in mathematics and science. In many instances, as in the following interview with a Massachusetts urban middle school teacher, teachers talked about the challenges of designing their curricula around teaching for understanding and simultaneously meeting the requirements of established curricular frameworks and formal standards:

Interviewer: If you want to think about your middle school group of colleagues, are there any issues that you're working on collectively involving the curriculum?
[...]
Teacher: We are working on how to manage to teach everything in our frameworks using inquiry, and interesting curricula that involve societal issues, all the strands ... of the national standards. How can we teach all of that using all of this—that type of thing—and fit it all in one year?

Teachers often reported that their work in their professional development groups prompted them to change their expectations for classroom interactions and to confront ambivalence about how best to teach, as in the following interview with a Massachusetts urban elementary school teacher:

What [the design collaborative] has ... [provided] for me is an outlet for those kinds of [big] questions and a place in which ... I can sort through the confusion because ... for me the whole question of confusion is related to my own personal ambivalence about how to teach.... So I become more aware of the dynamic of the confusion for me as a teacher [and] as a learner. Like I am more aware of what happens to me personally and how that leads me to doubt and all of that stuff.

New curricula and instructional styles often required the teachers that we interviewed to rethink their classroom assessments and develop new methods of evaluating student work, as these Wisconsin suburban elementary school and middle school teachers described:

Trying to assess what they know is completely different than the way we have done it in the past. You don't have the quiz. You don't have the end-of-unit test. And trying to find a meaningful way to show what they are doing to their parents without having to read ... three months of what they have written in a journal.... You are not going to teach everything. You are not going to cover everything that you have covered in the past. Pick some of those big ideas that you want to cover, and if you get all those, great, and if you don't, hopefully we will pick it up next year. (Elementary school teacher interview)

> I think that's what I struggled with last year, with giving some of the assess-
> ments ... I didn't know how to grade these tests. I felt like I'm putting
> these numbers down, I have no idea what these numbers mean.... I could
> see progress, and I could see which kids really got it, but I didn't know how
> to fairly put a percentage on it. (Middle school field notes)

As Fennema and Franke (1992) argued, when teachers are teaching to
develop their students' understanding of mathematics, their knowledge of
student reasoning becomes as important a form of teacher knowledge as
knowledge of the discipline and knowledge of how to teach the discipline
(Shulman, 1987). In spite of great advances in documenting the contours of
student reasoning involving early-arithmetic problem solving (Hiebert et
al., 1997), student thinking in other domains of science and mathematics is
seldom transparent; rather, it takes time and practice to interpret. This
posed a major challenge to many teachers whom we observed, including
this Wisconsin suburban middle school teacher:

> Every once in a while, there'll be a student up there, they'll be explaining
> the strategy, and I'm thinking, "You know, I don't think that works. I
> don't think you have it. I don't think you understand." But then there's al-
> ways in the back of my mind, I'm thinking, "Hmm. Maybe I'm just not,
> maybe it does work, but I'm just not seeing how it works." And I still strug-
> gle with, "OK. How do I handle this?" if I'm thinking to myself, "No, this
> doesn't work," but I'm really not sure. Because maybe I just don't under-
> stand what they're saying. (Field notes)

The Professional Development Group
as a Site for Managing Uncertainty

In many interviews, teachers talked about how discussions with their colleagues (researchers and other teachers) from the professional development communities helped them to focus their energies and develop
teaching for understanding more than they would have on their own. Discussions among community members took place at all times—during times
that had been set aside for meeting as well as during informal interactions
among teachers or between teachers and researchers. This Wisconsin suburban elementary teacher described how participation in the professional
development seminar influenced him positively, through providing a forum for developing ideas with colleagues:

> [The seminar] has definitely been a very positive influence on me in order
> to encourage me to discuss ideas with other people. It gives us the time
> and forces us to have these discussions in a sense. This particular year,
> having the resource days has been helpful because then we had specific

time where we focused in on something that we were personally working on. Last year ... I felt that I was kind of out there and wasn't really sure what I wanted to focus in on. Whereas this year I had more ideas about what I wanted to focus in on, so that I could become a better teacher. (Teacher interview)

The discussions occurring outside the professional development seminar also helped teachers to follow and respond to students' thinking more effectively by improving their own knowledge on a variety of topics. Here, a Wisconsin suburban elementary teacher relied on colleagues for help on the mathematics behind students' answers and advice on how to interpret students' responses:

I was looking at the kid's work, and I was having trouble understanding if the kids were right, because I really wasn't understanding the rule that I was coming up with.... So I was asking [two researchers] and a couple of other people, and it was basically kind of an algebra formula ... And then I just kind of sat down, and [they] helped me see the pattern, helped me understand the math behind it, and then I really needed to understand that before I could go further with my kids because I can kind of teach something with kind of half knowing what it is, but when they start talking to you ... you really have to know it. You really have to understand the [ideas] underneath of it. And I was blocked before in the past by not really understanding some of the stuff. And so it just kind of cleared itself up. And once I got that I was fine. I understood everything. It really helped me with that. (Teacher interview)

The content that teachers discussed covered the range of topics—curriculum, teaching, assessment, and student understanding. This Wisconsin urban high school mathematics teacher considered the value of depth versus breadth in terms of content coverage:

I mean, if I taught you in-depth, am I having a better prepared student rather than a well-rounded student? Who would be more prepared, I guess, is kind of my question to you. One who has learned more in depth or one who is well rounded? What would you say to that? (Field notes)

In addition, the professional development provided forums for teachers to make their questions explicit, so that they could recognize directions for growth. One Wisconsin suburban middle school teacher reported how, though she had learned so much about assessing student understanding, many questions remained:

I feel like I'm just starting to get a handle on how to assess student work, and I still have a lot of questions about how to do that. If I think about what

I learned in teaching math this year through the [design collaborative], it was getting a better handle at how I can make a student learn. So [at] some of our after-school sessions that we had I got some really good ideas from [a researcher] about looking at homework. I'm still working at trying to decide how we write a good assessment, and ... [two researchers] came in one day to talk about one of the assessments I had, and I got a lot of feedback, and that was really valuable to me. That's one I'm still struggling with ... how do we write a good assessment, how can we best give parents the feedback saying, "These are the skills your [children] need to continue to work on, these are the skills they feel comfortable with, this is what they're doing." (Teacher interview)

The Creation of New Technical Knowledge

The professional development groups provided settings where teachers could bring their beliefs to the surface and inquire within the group about the basis, either in their own experiences or within their shared values, that supported or disconfirmed those beliefs. Within the group, teachers could discuss how their ideas worked—or had failed to work—in their classrooms. By addressing these and other issues surrounding instructional uncertainty, the professional development groups provided an important social environment for developing and testing beliefs, and for creating new technical knowledge from empirically and socially validated beliefs. A Wisconsin suburban elementary school teacher described how participants in her design collaborative collectively developed an instructional tool to help students understand triangles:

When we did triangles, kids were stymied with one particular question on what happens to the adjacent side if this angle gets bigger, and [we were] trying to figure out how we can get them to understand that the side across from the angle will also get bigger and it is proportionate. And sitting down with people ... we all knew what we wanted the kids to understand, and we came up with this tool and what we had been working on, for it seemed like a week, all of a sudden we had this tool, and it worked. Within one class period, they were saying, "Oh, yeah, I see what is going on." ... And that was, that is something that would never have come up, and we probably could have scribbled in another week just talking about it. But without all of those people sitting down and talking ... that tool would have never, would have never happened. (Teacher interview)

SCHOOL SCIENCE ASSESSMENT:
THE CASE OF OBERON HIGH SCHOOL

The case of Oberon, the Wisconsin suburban high school site, provides additional evidence to illustrate our claims involving uncertainty. The Oberon research site included six of the seven teachers in the school's science department and several researchers. In a formal presentation to colleagues at the research center, the research team outlined issues that emerged for teachers in the design collaborative as they began to apply a framework for using synthetic models and scientific argumentation to develop students' science understanding:

> Helping teachers to use this framework in their teaching has proven to be challenging in a number of ways. First of all, it is immediately evident to teachers that they will teach less of their traditional content if they are to allow time for students to create and critique models in their classrooms. The teachers are frequently uneasy about this "trade-off," in part due to perceived pressure from standardized assessments and college entrance requirements and in part due to their own views on what material ought to be covered. Furthermore, there are a number of issues that arise when teachers attempt to use this framework in their classrooms:
>
> - Helping students create and defend their own models requires that the students be active learners. Therefore, teachers need to establish and communicate norms of student behavior with respect to group work, active participation, and regular preparation.
> - Teachers need to be able to use the framework to guide their interactions with students while the students are engaged in modeling.
> - Teachers need to design and implement authentic assessments—that is, assessments that require students to demonstrate knowledge in familiar formats and test their skills of argumentation in addition to their understanding of content.
> - Teachers need to continually assess the "invisible" learning outcomes related to argumentation. These argumentation skills are more difficult to assess than traditional content and are therefore less "visible" to teachers and students. Thus, they need to be regularly reinforced/emphasized. (Field notes)

The narrative that follows focuses on just one small facet of these larger challenges. Teachers were unsure about how to assess, interpret, and grade student performance on new assessments that were intended to reveal student understanding of complex scientific phenomena. Not only did the classroom assessments have to serve the teachers' instructional goals, they also had to allow students to demonstrate what they actually could do, and the grading scheme had to be accepted by all interested stakeholders (teachers, parents, and students). We discuss how the uncertainty about assessment became more salient, how the professional development community helped teachers to manage that uncertainty, how sharing was important and allowed the community to accomplish its goals, and how a new body of technical knowledge formed through the community's efforts.

Increased Salience of Uncertainty in Student Assessment

The teachers at Oberon High School came to understand that science teaching that emphasized modeling would require an approach to curriculum planning that emphasized flexibility and modification. They often talked about the difficulty of such planning. In particular, the issue of how to assess this type of student learning became a common theme in the design collaborative meetings. In one instance, a teacher declared, "I haven't a clue about how to give a knowledge and skills test for the modeling exercise" (Field notes). In another discussion, one researcher said that her interviews with students revealed how they had specifically used prior knowledge in figuring things out. In response, another researcher asked how a teacher might track something like that with many students in a way that was not too time consuming (Field notes). The group initially relied on students' journal writing to obtain information about student thinking. However, some teachers expressed frustration about the effectiveness of journals as assessment tools. This teacher pointed out that journals did not provide teachers with daily feedback from students:

> My biggest frustration … is the journals … Something needs to be done. There's not enough structure for a lot of these kids. They stare at a blank page and they're asked to write stuff down, and it's … just not there … And I have this very, very diffuse feel for how the kids are doing right now. And … for me it's really something that eats at my core 'cause I like to know on a daily basis how kids are doing. (Field notes)

The professional development community, of course, had developed end-of-unit tests that were designed to allow students to demonstrate their understanding of scientific content. When students did not perform as well as expected on one such exam, the teachers and researchers questioned whether their assessment reflected accurately what the students knew. One researcher suggested, "An important question is whether the problems are with the students' understanding, or how it's assessed. I'd be interested in taking a systematic look at these exams now" (Field notes). Specifically, teachers expressed dissatisfaction with the unit tests' failure to assess all of the student learning that had taken place as a result of the new curricular units that the group had developed. One teacher noted that "Thinking of what we were doing in the fall with assessments, I think we all felt like we had students learning in one way and then we were testing them in another way. That was very frustrating. That was a common thread for all of us, that we need to come up with something so that we all can feel better about it" (Teacher interview).

Another commented that

When it came time to give a test that was a written, and a very different task than using a three-dimensional model or explaining on a poster, a very different task, we found oftentimes when we know they understood the model, because they could show us with the models and the object, they didn't come up with correct responses on a written test. We weren't sure where the breakdown there was. How to write a two-dimensional, written ... test that will evaluate those tasks is a difficult question. That is something that we haven't resolved yet. (Teacher interview)

Beyond uncertainty about the usefulness of journals for revealing student reasoning and about test performance as a valid indicator of student competence, teachers expressed uncertainty about how to interpret student performance on assessments that drew on the modeling approach that they used in teaching. Were results due to the increased amount of time that they spent on each topic, to the new curriculum's content, to how they taught that content, or to some combination of all three? The following conversation took place between a researcher and two teachers in reference to a new curricular unit that the group had developed to teach phases of the moon:

Mr. N:	We spent a lot of time on the [phases of the moon], I think. Now I'm not saying that's bad ... you know, it was a couple of days, three days ... I can't remember, I'd have to look back ... and probably a little bit longer than that. I don't know if it, it seems to me that this idea of teaching modeling ... if it's that we're just spending more time in doing something versus less, how can I say this? ... At times I felt like we were doing the same thing, in class, over and over and over, and that's why the kids got it. Versus some other modeling idea. Does that make sense? ... What's something we've done recently ... the vector unit ... if we spent the same number of class hours doing just vectors as we did phases of the moon, I'm pretty sure the students would know vectors as well as they know phases of the moon. Does it make sense what I'm saying?
Researcher:	So you're trying to pinpoint a causal argument in some sense ... [...]
Mr. N:	Right. [...]
Researcher:	Do they understand this well because of this curriculum and instruction technique, or do they understand it well simply because we spent a lot more time emphasizing it?
Mr. N:	Exactly. And it's kind of going to what [another teacher] was saying. I'm not sure, you know. Granted we're giving them more time in class to build this model in their mind, um, but I'm not saying it's necessarily bad, don't get me wrong. I'm just—I'm trying to figure out what is the real ...

Ms. B: But I think part of that is the point … is I think that, by modeling, it forces you to look more in-depth at something as opposed to just passing it by. I mean I don't know if it's necessarily associated with just modeling or with anything, but I think kind of that's what's nice [emphasis] about it is instead of giving it a drive-by, um, you do spend time. I mean that's the whole.

[…]

Mr. N: Obviously, obviously, but my point is, at the expense of what? And I guess that's what I'm kind of thinking about. (Field notes)

In addition to concerns about gathering assessment data and how they might use it to plan instruction, teachers were also concerned about grading and assessment's more public uses. The following teacher expressed dissatisfaction with her existing assessment rubrics but acknowledged the challenges involved in devising better ones:

> The hardest thing is always the evaluation tools. For example, when a group presents their model, defends their model in front of their peers, it is hard for me to figure out how to grade that. How do you give that an A, B, C, or D? What do you add points for? What do you take off points for? It is hard to, you know, that is something that I always have a difficult time with. And that is something other teachers might have a rubric [for] that they find effective, but it seems like I am never satisfied with their rubrics, and I am not very good at improving them. It just seems hard to me. And it is a source of frustration for me. (Teacher interview)

Change and improvement take time. After working together for some years on their modeling approach to teaching science, the teachers began taking a longer view of their efforts. Such a view helped them to tolerate the questions, imperfections, and nagging doubts about technical problems and issues. This teacher discussed the time involved in changing practice and the importance of having patience:

> Look at [an outside researcher who worked with the group on assessment], she says it takes three years really to do a good job thinking about assessment for a program. And you know, we are used to fast food … and you can't do that. It has to, just like in a modeling, you revise things. And when you are doing a curriculum you have to always be open to, "Oop, this didn't work, why didn't it work, what should we change to make it work better?" We have to do that. (Teacher interview)

Managing Uncertainty Within a Professional Community

The Oberon teachers' uncertainty was not limited to their assessment practices; we found examples of their struggle to make sense of curriculum planning and to decide how they should teach, in light of their emphasis on modeling as a vehicle for promoting student learning of science with understanding. But we do not want readers to think that teachers experienced nothing but uncertainty. In fact, the entire professional development community celebrated many successes. What is more, teachers began to depend on each other for advice and to come up with mutually beneficial ideas. This teacher explained how her uncertainties made her want to interact more with her colleagues so that, in their collective work, they could design better solutions to the problems they encountered:

> I guess, in a sense, whenever you are around people who are knowledgeable and creative, I think there's a sense of doubt about your own capabilities. I couldn't be doing this all on my own, and then I think, "God, this person has such a great idea with that, and this person has such a great idea with that," and I didn't have that idea, it just kind of makes you realize how important it is to be around other people. And I say not having much confidence in your own abilities, it's not because you think that you are not intelligent or that you don't have the capability of being a good teacher. But just because you realize one person is not going to be as creative and dynamic as a lot of people working on the same problem. So I think that worries me, to go back to a situation where I might [be] more on my own and that I won't be doing as good of a job. (Teacher interview)

As noted, teachers at this site came to rely on the group to come up with solutions that were better than those that a lone teacher might derive. They also relied on the group to test ideas. This teacher spoke first about how frustrated he had been with using journals to obtain feedback from students, but then continued:

> I've got a few suggestions as to how to change that. But I don't know if this is an appropriate time to talk about it. But I would say ... I think the kids need sheets handed out to them with the questions already on them, that they can write on in their spaces.... We're going to take the time to print stuff out for them, photocopy them, and just give them full sheets, and they, we can hole-punch them, and they can keep their own little folder, and they can hand in individual sheets and things of that nature. They can build their journals that way and keep track of them. (Field notes)

Another teacher described her success in orally assessing student understanding. At first, a student demonstrated many misconceptions about the phases of the moon in relation to the earth and the sun. But the teacher had three-dimensional examples for him to look at, and eventually the student figured out the right answer by studying those visual representations:

> Well, what was really cool about the interview was he just kept looking at it, and I just kept slowly moving it around ... and he ultimately came to

the [solution and] ... he actually applied what was shining there. So I mean, on a quiz, if you're just gonna fill in an answer and you don't have something to manipulate, that might really be harder. But I don't know, it was a fascinating interview because at first I thought, "Oh my god, he doesn't remember anything," but then he just worked it through himself. And in the end, he had the most gorgeous picture of the crescent moon. (Field notes)

Members of the professional development community continued to build on one another's insights, thereby elaborating what was said earlier but also validating other teachers' experiences. In one design collaborative meeting, a teacher talked about how much she had learned from assessing her students' understandings by simply walking around the classroom and listening to them. This sparked a discussion about students' science misconceptions and how the group dynamics of a class can affect student thinking:

Ms. B: I feel that I know more about what they know than if you would have done it in a more traditional way. Because I, like you'd walk around and you hear them talking and, and I would just sometimes be flabbergasted as to what they were thinking. I'd be like, "I would've never in a million years guessed that that's really what you were thinking." You know, and you could really start to pinpoint where they were getting some of these ideas from. And ... what's interesting is some of the, one of the girls ... she said something that may have been told in elementary school, and I can't remember what it was, and she said, "But I don't ... understand where you guys are getting confused about the moon and shadows ... we were taught in elementary school that night means that the moon is out," ... and there was another one where they keep holding on to the shadows idea.

Ms. C: Yeah, because I brought that up so many times in class ... every day. And you always get one or two people that say the shadows.

[More talking about some of students' misconceptions that come up in class. They have to work at getting students to think about where they get their preconceptions. Often even after they have explained something over and over, the students still ask the same questions. Steve says they learn it better after they've worked it through on their own. Also the particular group dynamics of a particular class make a difference. Becky says that what she liked about how they've been setting things up is that they didn't have students model the whole time, but they alternated small group with large group activities throughout the class time.] (Field notes)

Because of these kinds of interactions, the teachers began to see formal and informal oral assessment as a possible supplement, if not an alternative, to journals. The group also provided a forum for questioning recommendations that the teachers received. When an assessment expert, who was helping the group to understand their assessment practices, proposed that teachers record their observations and evaluations about student understanding each day while walking around the classroom, one teacher reacted, "I could never do that ... or could I?" (Field notes). Another topic of uncertainty involved the grading of student work. The group discussed this problem, as it did others, both in the abstract and with concrete examples. For example, while the group was looking at a set of students' astronomy posters and discussing strategies for evaluating them, a researcher noted:

> It seems like you [as the student] can come to this task with an understanding of the original model, and then do something here where you don't have data–model match. Um, because this is a situation where they're sort of extending it, it's new, it's different, like they can explain seasonal effects at 23 degrees, but at 90 degrees they can't. And I guess, I mean I don't know if that's a useful distinction or not, but I think that it's slightly different than asking about what they understand about the original model. Does that make sense? (Field notes)

The group continued the discussion of specific student posters and how to evaluate them. A teacher "noticed that students made superficial connections in a lot of their writing without really clarifying what they mean" (Field notes). This observation led to more discussion until a researcher "pointed to a specific poster. She noted that here you don't have specific information about the links between model–data reasoning and understanding, and that's why this poster falls short" (Field notes).

Beyond its helping teachers to address, and in some cases solve, technical problems about assessment, the professional community provided teachers with emotional support that helped them to persist, even in the face of administrative indifference, if not opposition. Because such indifference can further erode teachers' confidence in what they are doing, peer support can be useful in dealing with the challenges of making curriculum and assessment fit with formal standards and frameworks. One teacher explained how she would advise another teacher who wanted to make similar changes:

> Try to have group support again, peer support, especially if you don't have administrative support, then you need [it]. See, I'm not sure we're [going to] have administrative support on this yet, I don't, it's just all talk. Once it comes down to curriculum and accountability with the state standards I don't know what's going to happen. So if you want to be stub-

born and do it anyway, then you need support, peer support. (Teacher interview)

Sharing and Collaborating: Group Mechanisms for Managing Uncertainty

In addition to using their communities to generate possible solutions for technical problems, the teachers recognized each other's contributions to their own personal efforts. After having expressed frustration about the journal assignment, Mr. N acknowledged Ms. B's help in coming up with ways for assessing students' work in the journals, telling her, "Without your input, I would have had no idea about how to grade anything from those journals" (Field notes). As the group's role in helping to resolve problems became clearer, teachers relied on "sharing." One teacher reflected how the professional development seminar paid increasing attention to issues of assessment after it became apparent that the existing assessments and grading techniques were not capturing what they had observed students learning:

> I think that [the group's regular meeting] is where it was brought out a lot. People were not satisfied in assigning grades, knowing where their students were at during the unit. So that was brought back to the big group and *shared* [italics added] and then discussed, and I think that is where it was pretty well agreed on, that we need to look at the assessment end of things in a more detailed matter, which we didn't spend time on during the summer. [We] kept saying, "We got to do assessment, we got to do assessment." (Teacher interview)

Another teacher noted how collaboration through the deprivatization of practice (see Newmann & Associates, 1996) and communication among group members were crucial in addressing concerns about assessment:

Teacher: They [other teachers] may be standing in the back [of the classroom] watching what you do, and you're thinking, "Oh god, I can't believe I did that," but they're also there, you know, "I've made these handouts or I've got this thing, what do you think of this if we pass this out to the kids? What do you think of that?" And you know they make a few changes and that, but, no, I think ... the support of the group in general is helpful.

Interviewer: When you tried out ideas from the professional collaborative in your classroom, or ... observed them being tried out in other classrooms, did you encounter any problems, or did you have any questions, and if so, what were they?

Teacher: Sure. Sure. Like I said, it was the first time that we were doing this with the freshmen. Um, there were some things that worked. Assessment was a real big problem for people. And you know, I say that in the positive sense, I mean, assessment is my weakest area, and I admit that. Um, but I think what's nice about it is we admitted that ... and we said, "What are we going to do?" And now we've got [the assessment expert] working with us on how we can assess differently. So I think assessment was a big problem. Um, but we're working on it. Other problems ... I think there were different personalities, and some people who had more of a share in a particular unit than others ... So I think that's one thing that we learned ... that you know if you're gonna have a group of people working on a course, they should all be involved in doing [all] the aspects of that course. So we learned that.

Creating Technical Knowledge

It is our claim that the professional development group created a body of technical knowledge around issues of assessment as they related to modeling and student understanding of scientific phenomena. This body of knowledge was created not just through the seminars, though these provided an important site for validating conjectures and exploring alternatives. Nor was knowledge created by teachers working alone in their classrooms in the stereotypical image of Lortie's (1975) school teacher. Rather, knowledge was created through the give and take involving many public and private sites. Teachers' classrooms served as sites for trying out ideas about ways to assess student understanding; student work became data that were discussed; private conversations among teachers and observations in one another's classrooms provided refinements for what teachers tried out in their own rooms; and the seminar provided a public forum for scrutiny of results, group discussion, discarding of failed experiments, and, eventually, adopting what worked. As in any good laboratory, public scrutiny and validation of results were integral to the creation of knowledge.

From our field notes, we found that newly developing technical knowledge was marked by the warrants that validated it: When speakers made claims, they often stated why they thought something was true. Among the new bits of knowledge created by this group was an agreement that visual props are invaluable for helping students show their understanding. One researcher commented that, "It's much easier to understand what students are talking about when they can show what they're talking about at the same [time], because that's how they're learning it in class" (Field notes). The group also agreed that in-class evaluations of student learning (i.e., assessments that were graded) should mirror how students had learned particular content. Another researcher observed, "Well, it seems like if that's how they're learning, their assessments should be more correlated with that … If they always get to talk about their model, and revise it with others, and use objects, I don't understand why the quiz should be such a different environment" (Field notes).

Hence, the presentation of scientific knowledge through poster sessions that the teachers had designed collaboratively became an acceptable vehicle for students to demonstrate their understanding. A teacher described the payoffs of these poster sessions during a design collaborative meeting:

> They not only let you be impressed with what they know, it lets them be impressed with what they know. Because when people start coming around that don't know this, and they start explaining it to them … And it was the same thing with the seniors. They were really pleased with [it], they seemed to value the whole nine weeks of class twice as much from that one hour experience, because they realized what they'd done and what they were able to show other people. (Field notes)

The group continued to discuss the utility of the poster sessions:

> In response to [a teacher's] question, "How well does this group represent students at large? Is this an above-average group?" the group discussed the individual students who did the presentations. The students represented a range of abilities. Some were above average, and some were below average in general. Not all who did good presentations are the best students. One student who did the poster session is severely learning disabled and has a really hard time on tests, and another has a lot of difficulties in class. They were very average as a group, representing a range … [Then the teacher reflected that] "This really underscores the idea that [there is] the need for appropriate assessments. You guys said that … they did pretty poorly in general on the written exam, right?" (Field notes).

The outside assessment specialist provided external validation of their accomplishments. She summarized how, in the process of creating the poster presentation activity and collaborating on a strategy for assessing it the group developed a set of criteria that clearly specified the desired student learning outcomes:

> I heard four major themes from your last conversation that seem to indicate that, intentionally or not, you already have some criteria that you've assumed as a group … There was a lot of "We want the science to be accurate, we want the subject matter to be accurate." … Collectively, you were really committed to understanding. And phrases you use are "understanding," "good thinking," "being able to write and think," "be able to

11.   UNCERTAINTY AND TECHNICAL KNOWLEDGE **273**

talk," "be able to," my phrase, not yours, "describe the model to a naïve audience" ... "They use the words, but we're not sure we know what the words mean." "Can they conceptualize, can they be concrete, can they integrate?" ... So there was a whole bunch of stuff about students' understanding. So it was more than just having accurate science content knowledge, but being able to show that there was accurate kinds of content knowledge that had been personalized in some way. There was also a considerable amount of conversation about what I would call extension. "I like to interview them because when I changed the parameters I could see if they really understand." ... And, "Could the kids, could they think on their feet?" ... Probably to me the most profound discussion that you had was one where you suddenly ask yourself, "Well, what were we expecting out of this assignment anyhow?" ... To me, this was exciting and profound, because this is one of the first steps about setting up [criteria] to be clear about what it is you want the students to be able to demonstrate by the assignments that you give them. And I was just kind of curious ... Have you ever had this discussion about student variability and student effort when you're giving a multiple-choice test? And why does a project bring this to the fore rather than a multiple-choice test? And I don't want an answer. (Field notes)

Ms. B responded with an observation about the gradual development of the assessment rubric:

It's interesting because, when we were doing our poster, we started off by just saying, "What are some categories that we're gonna look at, and then we're gonna comment on each one." And we, overall, put it into content versus presentation, kind of looking at those as two major divisions. And then within the content, we had, "How accurate, how complete, what's the level of detail, the clarity, how well-organized is your information?" Whereas in presentation we had, you know, "Is it easy to read, is it eye-catching, is it clever," you know, "creative, organized?" (Field notes)

CONCLUDING COMMENTS

We would agree with those who would claim that much of the uncertainty in teaching for understanding finds its roots in the relatively short time that this form of teaching, what Cohen (1988a, 1988b) called "adventuresome teaching," has been in development. As some practices become codified and are routinely adopted by teachers, we believe that the dilemma of uncertainty will decrease. Some might argue that teaching for understanding is an art form and that nothing about it can be codified or routinized. To this claim we would respond that teaching for understanding is like jazz. In order to improvise, the jazz master has routines that allow her or him to return to a theme at any time that is necessary. In addition, we would hypothesize that teachers who have been teaching for understanding for long periods of time are likely to possess well-organized and highly elaborated knowledge about student reasoning, curriculum, instruction, and assessment. That knowledge could provide an important source of professional practices that educators could codify.

Whether all of teaching for understanding can become codified in the form of technical knowledge, however, remains an open question. Even so, teaching for understanding makes explicit the relationship between how students reason and what teachers do next: It forces teachers to recognize how their success is intimately tied to their students' learning.

Regardless of the ultimate solution to the question of whether teaching for understanding can be codified, we believe that teachers' professional communities provide an important social mechanism for managing the uncertainties that currently are so pressing. Also, we believe that the work of these communities will become an important source of the knowledge base that will come to undergird teaching for understanding in mathematics and science.

Our results build on the work of others who study teacher change and development. Many researchers (e.g., Franke, Carpenter, Levi, & Fennema, 2001) have examined how individual teachers engage in long-term change and use their classrooms as places for the study of their students' reasoning as well as laboratories for their own practices. We extend this by documenting how collaboration can support this change.

These hypotheses and illustrative data provide a possible response to those who question the need for teachers to have common planning time, in which to address the pressing problems of teaching for understanding, and to those who would question the need to reduce teachers' isolation from one another. Teachers who work together are engaged in solving problems that are beyond the ken of individuals working alone. At the extremes, examples of individuals who achieve tremendous results through working as heroic isolates certainly exist. Our work, however, suggests that social interaction among teachers, in a professional community, plays a critical role in their growth and in the validation of their findings as knowledge.

AUTHOR NOTE

The data sources for this chapter are described in more detail in Gamoran et al. (2003). In order to maintain confidentiality, personal and school names have been changed to pseudonyms and minor personal details may have been changed. Research was conducted through the National Center for Improving Student Learning and Achievement in Mathematics and Science, supported by funds from the U.S. Department of Education, Office of

Educational Research and Improvement (Grant No. R305A60007). Findings and conclusions are those of the authors and do not necessarily reflect the views of the supporting agencies.

REFERENCES

Ball, D., & Rundquist, S. (1993). Collaboration as a context for joining teacher learning with learning about teaching. In D. K. Cohen, M. W. McLaughlin, & J. E. Talbert (Eds.), *Teaching for understanding: Challenges for policy and practice* (pp. 13–42). San Francisco: Jossey-Bass.

Cohen, D. (1988a). Teaching practice: Plus ça change ... In P. Jackson (Ed.), *Contributing to educational change: Perspectives on research and policy*. Berkeley, CA: McCutchan.

Cohen, D. (1988b). *Teaching practice: Plus ça change ...* (Issues Paper No. 88-3). East Lansing, MI: National Center for Research on Teacher Education, Michigan State University. [http://nctrl.msu.edu/http/ipapers/html/ip883.htm]

Fennema, E., & Franke, M. L. (1992). Teachers' knowledge and its impact. In D. Grouws (Ed.), *Handbook of research on mathematics teaching and learning* (pp. 147–164). New York: Macmillan.

Fennema, E., & Romberg, T. A. (Eds.). (1999). *Mathematics classrooms that promote understanding*. Mahwah, NJ: Lawrence Erlbaum Associates.

Franke, M. L., Carpenter, T. P., Levi, L., & Fennema, E. (2001). Capturing teachers' generative change: A follow-up study of professional development in mathematics. *American Educational Research Journal, 38*(3), 653–689.

Gamoran, A., Anderson, C. W., Quiroz, P. A., Secada, W. G., Williams, T., & Ashmann, S. (2003). *Transforming teaching in math and science: How schools and districts can support change*. New York: Teachers College Press.

Gamoran, A., Secada, W. G., & Marrett, C. B. (2000). The organizational context of teaching and learning: Changing theoretical perspectives. In M. T. Hallinan (Ed.), *Handbook of research in the sociology of education* (pp. 37–63). New York: Kluwer Academic/Plenum.

Hiebert, J., Carpenter, T. P., Fennema, E., Fuson, K., Human, P., Murray, H., et al. (1997). *Making sense: Teaching and learning mathematics with understanding*. Portsmouth, NH: Heinemann.

Lampert, M. (1986). Knowing, doing, and teaching multiplication. *Cognition and Instruction, 3*, 305–342.

Lampert, M. (1990). When the problem is not the question and the solution is not the answer: Mathematical knowing and teaching. *American Educational Research Journal, 27*(1), 29–64.

Lortie, D. (1975). *Schoolteacher: A sociological study*. Chicago: University of Chicago Press.

Marks, H. M., Secada, W. G., & Doane, K. (1996). Support for student achievement. In F. M. Newmann & Associates (Eds.), *Authentic achievement: Restructuring schools for intellectual quality* (pp. 209–227). San Francisco: Jossey-Bass.

Newmann, F. M., & Associates. (1996). *Authentic achievement: Restructuring schools for intellectual quality*. San Francisco: Jossey-Bass.

Rowan, B. (1990). Commitment and control: Alternative strategies for the organizational design of schools. *Review of Research in Education, 16,* 353–389.

Rowan, B., Raudenbush, S. W., & Cheong, Y. F. (1993). Teaching as a nonroutine task: Implications for the management of schools. *Educational Administration Quarterly, 29*(4), 479–500.

Shulman, L. S. (1987). Knowledge and teaching: Foundations of the new reform. *Harvard Educational Review, 57*(1), 1–22.

CROSS-CUTTING STUDIES

Chapter 12

Research in Assessment Practices

Jan de Lange
Freudenthal Institute, University of Utrecht

Thomas A. Romberg
University of Wisconsin–Madison

Student understanding of conceptual knowledge in any field involves growth in knowledge and the ability to use that knowledge both creatively and routinely in solving a variety of problems encountered in the course of life. The question of how one judges student performance or student progress toward understanding, however, remains both problematic and controversial.

The goal of the reform assessment, as stated by the National Council of Teachers of Mathematics (NCTM) in its *Assessment Standards for School Mathematics* (1995), is that teachers should "monitor students' progress to understand and to document each student's growth in relation to mathematical [and scientific] goals and to provide students with relevant and useful feedback about their work and progress" (p. 29).

TRADITIONAL ASSESSMENT

In traditional assessment, teachers generally have access to three sources to judge their students' knowledge of mathematics and science:

- Externally developed (and often mandated) tests, such as standardized tests.

- Curriculum-based paper-and-pencil quizzes and tests (administered primarily to assign a grade).
- Informal evidence gathered from interacting with their students.

Unfortunately, traditional sources of information alone neither sufficiently nor efficiently assess student understanding. Too often, as well, the reporting of student scores, particularly on externally mandated tests, is far from timely and, therefore, furnishes little information to guide instruction or to provide feedback to students.

External Tests

The use of external testing as a policy tool is not new. In the 1840s, Horace Mann, then secretary of the Massachusetts Board of Education and editor of the *Common School Journal*, concluded (among other comments) that the new written examinations being instituted in the Boston schools were superior to the old oral tests in that the tests prevent the "officious interference" of the teacher; they "determine, beyond appeal or gainsaying, whether the pupils have been faithfully and competently taught," and they enable all to appraise the ease or difficulty of the questions (Gerberich, Greene, & Jorgenson, 1953, p. 22).

The ideas about objective testing expressed by Mann remain central tenets for policymakers: A teacher's judgments about any student's progress is still suspect, the test items are assumed to be representative of what has been taught, and the difficulty of each item is considered to be apparent.

Mann in later issues of the *Common School Journal*, however, commented that the examinations used were inadequate in that they could capture only some aspects of "what students had been taught." The instruments now being used are still inadequate, and the information derived is of little use to teachers when planning instruction.

Standardized tests are currently a yearly ritual in U.S. public schools. Even though these tests often do not relate well to the desired content of classroom instruction, they continue to have an impact on instruction in that teachers take classroom time to prepare students to take the tests and often adapt what they teach to the content of the external test (Kilpatrick, Swafford, & Findell, 2001). The strength of such tests is that they do what they were designed to do fairly well. They are relatively easy to develop, are inexpensive and convenient to administer and score, and yield comprehensible, if generally inadequate, results. The primary weakness of such tests mirrors Mann's argument in that they fail to capture much of what students have been taught or whether they are able to apply that content knowledge in new problem situations.

Curriculum-Based Quizzes and Tests

Mathematics teachers have traditionally monitored their students' progress by giving quizzes and chapter tests, scoring and counting the number of correct answers, and periodically summarizing student performance in terms of a letter grade (Stiggins, Conklin, & Bridgeford, 1986). Haertel (1986) estimated that such tests occupied about 15% of students' time in secondary schools, yet an 1982 analysis of 8,800 test questions in 12 grade and subject combinations (elementary to high school) showed that almost 80% of all questions were at the "knowledge" level (focusing on, e.g., recall, identification, description) in Bloom's Taxonomy (Fleming & Chambers, 1983). Similarly, Senk, Beckmann, and Thompson (1997) found in their study of 19 mathematics teachers, that the teachers used tests and quizzes consisting primarily of low-level tasks and based 77% of the students' grades on information from them. In a case study of four teachers, Smith (2000) found that teachers often focused solely on completion of tasks rather than on the quality of answers to such tasks. Shafer (1996) noted that the assessments of student performance given by the teachers involved in her study usually were related to short-term objectives (e.g., How did students do on this lesson, unit?), not long-term goals (e.g., How are students progressing toward understanding?). Many researchers have also noted that teachers use the collection of evidence from tests and quizzes not to inform instruction to meet the learning needs of their students, but primarily to provide evidence for the grades they assign. Unfortunately, as Wilson (1993) noted, "What gets graded is what gets valued [by students]" (p. 412). Students in her study paid little attention to *ungraded* homework, self-evaluations, higher order thinking questions, and writing assignments.

Grading Practices

Grading practices in the United States have long followed certain customs and traditions. Beginning at some point in elementary school and continuing through high school, teachers have reported their evaluations of students' work by giving them letter grades (A, B, C, D, or F). There are certain assumptions that the public makes about this practice. One of those is that, in general, grades are based on the comparison of students with other students, such that the distribution of grades follows a bell-shaped curve; that is, there should be more Cs than any other grade, and fewer As and Fs. Another assumption is that what these grades describe is student achievement: The grade of C usually connotes average achievement; a grade of A stands for superior work. There is also a tacit assumption that some degree of comparability exists between the grades given by a teacher at one school and those given by a teacher at another. Naturally, there are exceptions to these practices and exceptions to the assumptions, but in general these are the practices accepted as "normal" in most U.S. schools.

Traditional grading practices have endured primarily because they are convenient. They have provided teachers and schools with a seemingly simple way to communicate to students and the public some information about what students know in an easily recognizable form. The truth is that this system is based on a set of assumptions that do not bear up under scrutiny:

- The collapsing of evidence about student performance into a single letter or number is not accurate (or fair). Assessment based on multiple sources of evidence yields rich information about what mathematics students know, what mathematics they can do, and their disposition toward mathematics or science. Collapsing evidence into one score should be used only with caution and awareness that much information is lost during the process (Evans, 1976).
- Grades are too often based on the completion of work, rather than on its quality. As Wilson (1993) and Smith (2000) noted, this can motivate students to complete assignments, but focusing on the completion of tasks rather than the quality of the work fails to "provide feedback to students about specifically where they have done well, where they need improvement, and how to modify their performance in those areas needing improvement" (Smith, 2000, p. 181).
- Most teachers feel that effort, in addition to achievement, should be incorporated into grades (Frary, Cross, & Weber, 1993; Stiggins, Frisbie, & Griswold, 1989), and many do not assess higher order thinking skills (Stiggins & Conklin, 1992).
- Research shows that grades are not comparable across teachers or schools (Natriello, 1992). There is no common approach to the way information is aggregated by teachers to derive a grade, although there does seem to be stability in the grading done by individual teachers, even when they assess student work in different subjects (Stiggins & Conklin, 1992). But the assumption that an A in one class or even in one school means the same thing as an A in another class with another teacher at a different school cannot be supported.

Most importantly, present practice does not serve students well because it ignores the rich and varied information about what students know and can do. Coming up with alternative models for grading requires teachers to reexamine some of their basic assumptions about assessment and the ways information gathered about students is synthesized and reported. Undoubtedly, teachers will continue to use quizzes and chapter (or unit) tests to assist in making judgments of student progress, but care needs to be taken to ensure that these quizzes and tests are aligned with content and instructional goals and that the grades given reflect student achievement and understanding rather than primarily their completion of work assigned.

Informal Evidence

Teachers commonly collect information from classroom interactions, including observations, samples of student work, and student self-evaluations. The utility of such information is that it can support instructional decision making and the work of teachers as they monitor student progress. But, as Bachor and Anderson found in their 1994 study, although reform teachers at four grades gathered a wide variety of such information, they were not clear about what, why, and how they were to use the information. Because student strategies and responses to tasks varied, teachers also had a hard time making reasonable judgments about students' performances. The teachers "lacked an interpretive framework to make sense of the assessment evidence collected" (p. 91).

Other researchers have noted that information from informal sources was rarely used for grading students. Graue and Smith (1996), in their examination of the assessment beliefs and practices of four Grade 6 teachers implementing reform materials for the first time, reported that although teachers realized that their past practices were inconsistent with their instructional goals, they did not worry about any discrepancy between what they observed and the marks they gave on tests. Clearly, teachers need a framework for collecting, summarizing, and evaluating informal evidence, but we emphasize, as Webb (2004) noted, that teachers are more inclined to devote the time necessary to reform their assessment practices when they see its potential for improving the learning and achievement of their students.

STUDIES OF REFORM ASSESSMENT

The Assessment and Achievement Study Group and the associated study, Research on Assessment Practices (both studies based in the National Center for Improving Student Learning and Achievement [NCISLA], were designed to document how teachers monitor each student's progress toward important mathematical and scientific goals, to identify the problems and impediments that teachers face as they attempt to implement assessments consistent with the reform perspective about teaching for understanding, and to design professional development that focuses on these issues. Also of consideration in preparing our materials and professional development was the need for any assessment system we developed to be broad enough to allow the tracking of the growth of individual students. In this area, the design task became threefold: specifying details of what progression in a given domain of mathematics or science involved, determining the kind of information useful in helping a teacher determine a student's progress at a given point in time, and determining how such growth could be aggregated and reported.

Instead of reinforcing traditional ways of judging student performance in mathematics and science (e.g., scores on external tests, data from short quizzes), researchers collaborated with teachers to create assessment materials and professional development that would support teachers as they worked to develop reform assessment practices in the classroom. We were aware that such changes would be difficult. As Westbury (1980) argued, the change involved not only the adoption of new techniques (in our study, in assessment), but the abandonment of safe, customary practices. The deep structures of formal and informal institutional apparatus, procedures, forms, and rituals tend to preserve the status quo, frustrating efforts at assessment reform. Just as students sometimes have difficulty in learning because of their adherence to previous (mis)conceptions, so also do institutional (and belief) structures carry intellectual baggage that impedes change.

Theoretical Background

The standards model underlying our research (adapted from Taylor, 1994) focuses on the characteristics of student performances and exhibitions. Instead of defining objectives in terms of what students know in a particular domain of content, educators define the complex performances and processes that are authentic to that discipline. The emphasis is on the quality of what students can do and how that reflects what they know. These standards for quality, criteria for expectations, and exemplars of student work are to be shared with students, teachers, and parents to promote a clear understanding of the educational goals and how they will be assessed. This model also requires professional judgment in the analysis of student performances. In mathematics and science, the knowledge required for analyzing student performances includes knowledge of the content domain, understanding of students' thinking, awareness of possible solution strategies, and developmentally appropriate expectations for performance quality. In our research, we also relied heavily on the Dutch Realistic Mathematics Education (RME) approach to mathematics instruction, which reflects Taylor's (1994) standards model and was the theoretical base for the assessment materials developed for the middle school reform curriculum, *Mathematics in Context* (National Center for Research in Mathematical Sciences Education & Freudenthal Institute, 1997–1998).

The Dutch RME approach to assessment, which evolved over a period of nearly 20 years, is closely aligned with instruction and is seen as part of daily instructional practice (van den Heuvel-Panhuizen, 1996). Rather than including just tasks that mimic the content covered, the Dutch include open tasks, the solving of which requires students to relate concepts and procedures and to use what they have learned to solve nonroutine problems. The tasks are grounded in the three levels of thinking developed by Dutch scholars (see Fig. 12.1). Level I thinking involves performing specific calculations, solving a given equation, or reproducing memorized facts (items often similar to those used in standardized tests). Level II thinking requires integrating information, making connections within and across mathematical domains, and solving nonroutine problems. Level III thinking involves recognizing and extracting the mathematics in a situation, and using that mathematics to solve problems by analyzing data, developing models and strategies, and making mathematical arguments and generalizations. The Level III tasks are more complex, in that students extend what they have learned as they explore new interrelated tasks. (For details about the variety of tasks used, see de Lange, 1987, 1995.)

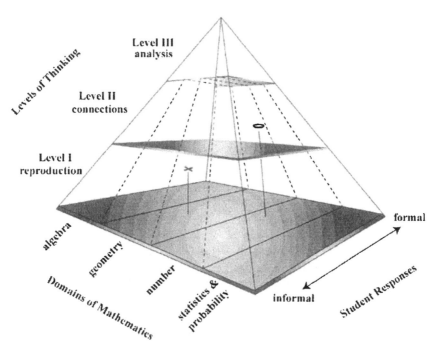

FIG. 12.1 Assessment pyramid. *Note.* Adapted from "Enquiry Project: Assessing Realistic Mathematics Education," by T. Dekker, 1997, unpublished manuscript, Freudenthal Institute, Utrecht, Netherlands; figure reprinted in "Mathematics Education and Assessment," by H. Verhage & J. de Lange, 1997, *Pythagoras, 42,* pp. 14–20.

Good assessment should provide reliable evidence of what a student is able to do in a given domain at a point in time. Ascertaining how many artifacts of the domain a student can identify is not sufficient. In the RME approach, assessment is designed to focus on the ways students identify and use artifacts to solve increasingly complex tasks as a student has additional mathematical experiences (often over the course of years). In our materials, therefore, assessment is intended not solely as a way to report student achievement at a given point in time, but also to function as a mathematical experience that supports student learning, extending student knowledge through exploration of nonroutine applications and enabling the teacher to provide feedback to students on ways to strengthen individual weaknesses as well as enhance to understanding.

Research on Assessment Practices (RAP) Project

The RAP project focused on research related to ways to help teachers document the progress of individual students from informal to formal understanding of concepts and procedures in both mathematics and science. Data were gathered on teachers' assessment practices at three sites through classroom observations, additional teacher interviews, and collection of student work. At the two mathematics sites (Verona, WI, and Providence, RI), staff monitored teachers in the process of implementing the reform mathematics materials *Mathematics in Context* (MiC; National Center for Research in Mathematical Sciences Education & Freudenthal Institute, 1997–1998). The work at the third site (Monona, WI) was done in collaboration with the NCISLA High School Design Collaborative.

The purpose of this study was to investigate how teachers in schools, some with diverse student populations, implemented a reform-based curriculum. The RAP team at two sites (Providence and Verona) assisted teachers in targeting the "big ideas" in mathematics (Steen, 1990) and offered assistance in documenting student growth over time. With classroom-based support and collaboration with RAP staff, teachers used tasks embedded within the curricula and worked to balance tasks in order to assess different levels of reasoning (for examples of the work at the Providence site, see Wijers, 2004). But creating tests that reflected reform goals was not easy, nor was it easy for teachers to implement new types of assessments without strong support. In using the new tasks, for example, teachers tended to give instructions such as "Explain your answer" or "Show your work." Given no benchmarks for justification and explanation, student work rarely met teachers' expectations. Similarly, teachers had difficulty judging the quality of student solutions and products.

The study of science teachers' assessment practices initially included only high school science classrooms in which students were invited to construct and revise their own "explanatory models" based on situations grounded in Mendelian genetics and evolutionary biology, but grew to include classrooms exploring Earth-Moon-Sun astronomy. Teachers at these sites were able to fully integrate assessment into the instructional system. Students learned to create, revise, and evaluate conceptual models. They worked in teams to explore their data and models, create poster sessions that explained their work, justify their findings to the class, and write up their findings. The assessment criteria were defined and negotiated in class, and the quality of work was often negotiated by the class and the teacher together. The teachers worked closely with Center researchers on all aspects of the curricula and assessment practices they developed and implemented.

Again we note that implementing reform materials and tasks—without appropriate professional development—is unlikely to be sufficient to meet reform goals.

REFORM ASSESSMENT: TASKS AND CLASSROOM PRACTICES

In the RAP project, we found that teachers needed a conceptual framework to ground their understanding of student growth in relation to the reform goals for school mathematics, and they needed examples of assessment tasks and clear guidelines for scoring and grading related to the reform goals. In fact, as Feijs and de Lange (2004) argued, most teachers need to become at least passive experts in classroom assessment tasks so they can adequately judge the quality of the tasks they select, adapt, and create. Such expertise should include understanding the role of the problem context, judging whether the task format fits the goal of the assessment, judging the appropriate level of formality (i.e., informal, preformal, formal), and determining the level of mathematical thinking involved in the solution of an assessment problem.

In response to these needs for information, de Lange (1999) and the Freudenthal Institute (FI) staff developed a *Framework for Classroom Assessment in Mathematics*, which reflects both RME's approach to assessment and NCTM's (1995) assessment standards. The intent of this framework is to generalize what had been learned about classroom assessment into a set of principles, with examples, so that the ideas could be used by teachers at all levels using any curricular materials. To complement this framework, the FI staff produced the *Great Assessment Problems* (GAP) book (Dekker & Querelle, 2002). This book, now available on the Internet, provides teachers with a variety of assessment tasks and takes a closer look at the pitfalls that might be encountered as teachers design, score, grade tests, and give feedback to their students.

Reform Assessment Tasks

The primary purpose of a classroom assessment system that reflects the standards-based reform movement is to improve learning. With that in mind, in this section we illustrate the kinds of tasks available to teachers by contrasting the difference in information about learning in the responses to a multiple-choice question and an open task, giving examples of tasks at each of the three levels in the RME pyramid, and discussing the use of group "investigations."

To examine the difference between a multiple-choice task and an open single-answer task, we gave two groups of students (in the same class) a 10-question test based on material from the GAP book (Dekker & Querelle, 2002). One group of students got multiple-choice questions; the other group got similar questions in an open single-answer format (see Fig. 12.2). After assessing the students' answers, staff summarized the differences between formats and the relative advantages and disadvantages of their use (see Table 12.1). Dekker and Querelle (2002) noted, however, that both formats are appropriate for problems on RME Level I.

In contrast, Level II and Level III problems almost always require an open format because evidence about a student's process of solving the problem is needed in addition to the product of their thinking. As Dekker and Querelle (2002) noted, "Answers to such open problems reveal more than just the ability to get the right answer. Students are expected to provide clear reasoning in an efficient and not too elaborate way" (GAP book, p. 8: 4). In Fig 12.3, the three correct answers are justified in quite different ways. The first explanation is a generalized statement, the second is based on an efficient calculation, and the final uses an example to justify the answer. The kind of justification reveals how different students reason about a problem.

Still another item format discussed in the GAP book (Dekker & Querelle, 2002) involves multiple-question items ("super-items"). Because complex problems are often posed within a context and it takes time for students to understand the context and the mathematics needed to address the problem, more than one question is posed to help the student solve the problem. The first question in such items is usually an introductory Level I question, which is followed by one or more questions at Levels I or II and ends with a question at Level III. In super-items, the problem situation needs to elicit different levels of response (by including some ambiguity, excess information, or a need to make an assumption), ask more than one question with respect to the initial problem situation, and suggest a reasonable scoring procedure, or rubric, for judging differences in responses. For example, the super-item shown in Fig. 12.4 requires proportional reasoning and focuses on understanding of both measurement and scale. To answer the first question in this item, students need only measure the spider using a centi-

A small bottle contains 200 milliliters of pure alcohol. Sandra forgets to close the bottle, and the alcohol starts to evaporate.

During the first day, 30% of the alcohol evaporates.

During the second day, 60% of *the amount still present* evaporates.

How much of the alcohol is left in the bottle after two days?

(A) 20 milliliters (B) 36 milliliters (C) 56 milliliters (D) 84 milliliters

Answer: (C)

Possible student work:

30% of 200 = 60
60% of 140 = 84
So 140 – 84 = 56 milliliters are left.

An example of an incorrect answer: 60% of 60 is 36

What is left is 200 – 60 – 36 = 104 millimeters.

Age: 14, 15   **Level:** I

Content: Quantitative reasoning, number

Context: Relevant

Situation: Daily life

NCTM: Understand meanings of operations and how they relate to one another

Note: A student in the multiple-choice group who did the same computation and reached this incorrect answer would not find his answer in (A), (B), (C), or (D) and would know that it could not be the right answer.

FIG. 12.2   Problem given in both multiple-choice and open formats.
Note. NCTM = National Council of Teachers of Mathematics. From *Great Assessment Problems*, by T. Dekker and N. Querelle, 2002, unpublished manuscript, Freudenthal Institute, Utrecht, Netherlands, chapter 8, p. 2.

meter ruler. To answer the second question, however, they need to measure the distance between 4 and 5 on the picture and estimate the scale. The third question then asks them to use that scale to estimate the size of the nail of an index finger drawn to scale into the picture.

TABLE 12.1

**Comparison of Relative Advantages and Disadvantages of Multiple-Choice
and Single-Answer Item Formats**

	Multiple-Choice Format	*Single-Answer (Open) Question*
Relative Advantages	*For the student:* • Can still get the right answer even if not using proper measurements. • If a small mistake is made in calculations, the answer obtained is not likely to be one of the choices. The student knows something must be wrong. • Even if a student does not have a clue about how to solve the problem, he or she might still guess the right answer.	*For the student:* • Partial credit for any part of the answer that is correct. • Not "misled" by wrong answers (i.e., no "multiple choice" to compare). • No temptation to pick an answer without doing some calculation.
	For the teacher: • Quick and easy to score.	*For the teacher:* • A totally unexpected answer provides evidence of a wrong strategy used by the student.
Relative Disadvantages	*For the student:* • If a student doesn't know the meaning of "a good 12%" or doesn't think properly about rounding off, he or she could easily pick the wrong answer.	
	For the teacher: • If a student notes an answer but shows no work, he or she could simply be guessing.	*For the teacher:* • Scoring takes more time.

Note. Adapted from *Great Assessment Problems*, by T. Dekker & N. Querelle, 2002, unpublished manuscript, Freudenthal Institute, Utrecht, Netherlands, chap. 8, p. 2.

In preparing open items, special care should be taken with the wording of the problem. The introductory text of a context problem and the questions themselves should be "translated" into student words if necessary, using short sentences and words that are not too complex. Wijers (2004) and Feijs and de Lange (2004) found that teachers had difficulties in de-

At a bakery, the price of chocolate silk pie is increased by 10%. Now fewer people

buy this pie. The baker decides to lower the new price by 5%.

He says: "Compared to the original price, the pie now costs 5% more than in the beginning."

Is the baker right? Give mathematical reasons to support your answer.

Possible student answers:

♦   The baker is wrong. The 10% increase is calculated from a smaller amount of money

than the 5% decrease. So in the end, the result is less than 5%.

♦ $100\% + 10\% = 110\%$

$0.95 \times 110 = 104.5$

The baker is wrong; the price is 4.5% more.

♦   I assume the price of the pastry was $3 originally.

$1.10 \times \$3 = \$3.30$ (10% increase)

$0.95 \times \$3.30 = \$3.135$

$1.05 \times \$3 = \$3.15$

The baker is wrong.

Age: 14, 15   **Level:** II

Content: Quantitative reasoning   **Context:** Relevant   **Situation:** Daily life

NCTM: Understand meanings of operations and how they relate to one another

FIG. 12.3   Level II open format problem.
Note. NCTM = National Council of Teachers of Mathematics. From *Great Assessment Problems,* by T. Dekker and N. Querelle, 2002, unpublished manuscript, Freudenthal Institute, Utrecht, Netherlands, chapter 8, p. 4.

signing assessment tasks, especially those that included Level II and Level III questions.

To facilitate the assessment of complex cognitive skills during instruction related to Level II and Level III goals, van Reeuwijk and Wijers (2004) identified complex problems that could be used as investigations—com-

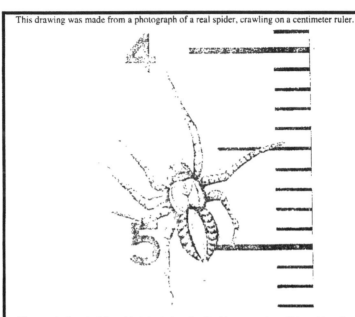

This drawing was made from a photograph of a real spider, crawling on a centimeter ruler.

a. Measure the length of the spider's body from head (with antennae) to tail, but without its seven legs. Use a centimeter ruler and give your answer in whole centimeters.

b. To what scale is this drawing made? Show your work.

Imagine the nail of your index finger is on the photograph this drawing was based on. Draw your nail to scale in the picture above. Explain what you did to make sure your nail was drawn to the right size. (**Note:** If young students have difficulty explaining writing down their reasoning, they can give oral explanations.)

Answers

(a) Length of the spider in the drawing is 5 (millimeters).

(b) The distance between 4 and 5 in the drawing measures 7.3 centimeters. In reality this length represents 1 centimeter exactly. So the scale of the drawing is 1 : 7.3.

(c) I measured the dimensions of my index fingernail and multiplied by 7.3 (or 7). Then I made a drawing of my nail according to these measurements.

Age: 10, 11   **Level:** Question a—I; Question b—II; Question c—III   **Context:** Relevant

Content: Quantitative reasoning, number sense, measurements   **Situation:** Daily (school) life

NCTM measurement standard for Grades 6–8: Apply appropriate techniques, tools and formulas to determine measurements

FIG. 12.4   Super-item with questions at Levels I, II, and III.
Note. NCTM = National Council of Teachers of Mathematics. From *Great Assessment Problems*, by T. Dekker and N. Querelle, 2002, unpublished manuscript, Freudenthal Institute, Utrecht, Netherlands chapter 8, p. 8.

plex tasks that involve mathematics from several content domains, take time (often days) to complete, and are often done in groups. Van Reeuwijk and Wijers gave one such problem, "Floor Covering" (Fig. 12.5), from the MiC unit *Reallotment* (Gravemeijer, Pligge, & Clarke, 1998) to students in project classrooms in Providence, RI. The task involves strategies for covering an area under specific constraints, then calculating the costs of area coverage using different materials. In one class, the students initially had difficulties visualizing the situation and could not think of a way to use carpet from a roll that was only 4 yards wide to cover a floor with dimensions of 6 × 14 yards. The teacher decided to provide more information as well as suggestions on how to start work on the problem. On grid paper, she drew the floor as 6 × 14 squares. Then, from another piece of grid paper, she cut one long strip four squares wide and rolled it up like a roll of carpet. She asked how much she would have to cut off this roll to cover the floor. The first suggestion was to cut a strip 14 yards long. After a discussion of how to cover the rest, one student suggested she cut four pieces, each 2 yards long, and put them on the remaining space that measured 2 × 14 yards. This filled the remaining area, with only 2 square yards wasted. Another student came up with the idea of cutting pieces 6 yards long. Four 6 × 4 pieces would cover the floor, leaving a small strip of 2 × 6 yards as waste.

This example illustrates the features of investigations and the difficulties in designing, administering, and scoring such tasks. Additionally, van Reeuwijk and Wijers (2004) demonstrated that students need to learn how to solve such complex problems by designing a plan, trying different strategies, and being aware of varied expectations for such problems.

Formative Assessment

From this work with teachers, we found that formative (or classroom) assessment must be seen by teachers as an integral part of instruction.

Two Verona teachers, Ann Fredrickson and Terri Her, reported that over time they developed a more comprehensive view of assessment and that they began regularly to use a wider range of assessment strategies (Fredrickson & Ford, 2004; Her & Webb, 2004). Their increased attention to student learning (via assessment) helped them see the relations among content, instruction, and the evolution of student understanding as students progressed from informal to formal reasoning in mathematical domains. Terri Her, an experienced sixth-grade teacher, described both the process she went through to change her assessment system and the product of those changes (Her & Webb, 2004). In the following excerpt, she described the features of her new assessment system:

> I combined these components to produce a system that met the following goals for my latest method of assessment:

Floor Covering

A lobby of a new hotel is 14 yards long and 6 yards wide. It needs some type of floor covering

6 yards

14 yards

There are four options:

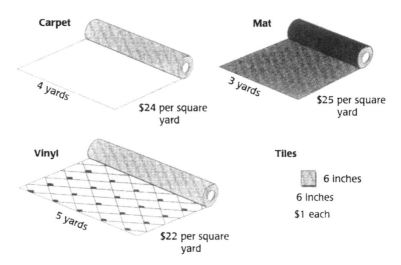

Carpet

4 yards

$24 per square yard

Mat

3 yards

$25 per square yard

Vinyl

5 yards

$22 per square yard

Tiles

6 inches

6 inches

$1 each

You are the salesperson for the floor-covering company. You have been asked by the hotel manager to present the total cost of each option along with your recommendation and your reasons for it. Write a report that explains your choice for the best floor covering for the hotel lobby.

FIG. 12.5   Floor covering investigation.
Note. From "Reallotment" (p. 31), by K. Gravemeijer, M. Pligge, & B. Clarke, 1998, in *Mathematics in Context*, by National Center for Achieving Student Learning and Achievement in Mathematics and Science and the Freudenthal Institute (Eds.), 1997–1998, Chicago, IL: Encyclopædia Britannica. 2. Copyright 1998 by Encyclopædia Britannica. Reprinted with permission.

1. Identify objectives for three reasoning levels, drawn from standards and unit content, to provide appropriate challenge for a wide range of student abilities.

2. Allow for both informal and formal assessment results to be recorded separately for each objective, using marks for "not yet," "in progress," or "demonstrated."

3. Permit students to make multiple attempts at improving and demonstrating understanding; do not give definitive marks until several opportunities have been offered.

4. Base letter grades on a performance rubric, not percentages.

5. Empower students and parents by making the content objectives as well as the marking and grading process explicit. (Her & Webb, 2004, p. 207)

Ms. Her went on to explain that when she started to use this system,

> I notified parents of the new grading system and explained to students the rationale behind my decision. Most students were interested in giving the new system a try. They enjoyed seeing up front the concepts they would be investigating and the process for determining their grades was now demystified. Some students were motivated to set personal goals and familiarize themselves with the requirements for earning the grade that they desired. Each student was given a copy of the unit objectives and the grading rubric, so students could keep track of their progress throughout the unit (goals they had demonstrated, what remained yet to be demonstrated, and their current grade). (pp. 209–210)

She summarized her findings under several categories. Under "increased student involvement, accountability, and self-advocacy" for example, she found the following:

> One immediate benefit in shifting from an emphasis on grades to an emphasis on assessing understanding was the impact it had on student communication and involvement. The student record forms, progress reports, and class conversations about assessment played a major role in increasing student fluency in the language of mathematics so that students were able to communicate the specific skills that they were learning and the concepts that they did not understand. They also were able to communicate what they were learning to parents and others. Even without the benefit of the forms in front of them, many students were able to articulate a clear understanding of and appreciation for the assessment process.... Student questions like "What is my percentage today?" or "How many extra credit points do I need to get an A?" were replaced with, "When can I demonstrate that I now can subtract fractions with unlike denominators?" or "How can I extend my thinking to go beyond the basic skills?" Parents stopped calling to haggle over points for lateness or lower-than-expected test scores. Students started advocating for themselves more, and they wanted to learn and demonstrate specific skills. As

one student remarked, "We are more concentrated on learning and not getting a good grade." (pp. 212–213)

And with respect to "raising the academic bar," she noted:

> Once I determined the three levels of reasoning I wanted to address, I was obligated to plan tasks and activities that could elicit such thinking. I was required, as a result, to raise my own awareness and knowledge of math content so that I could offer challenging analysis-level extension tasks to students who wanted to press beyond basic grade-level requirements. Students continuously rose to meet these challenges and attempted and achieved more than ever before. (p. 214)

The benefits of changing her assessment system were apparent to Ms. Her, but the process of change took more than two years of involvement in the implementation of a reform curriculum and the support of administrators, other teachers, and research staff.

BEYOND TASKS AND TECHNIQUES

Ms. Her's testimony is important, but in order to organize a classroom that enables students to engage with instruction that leads to understanding and to validly judge their progress toward understanding in a domain, several dimensions need to be considered in addition to the use of curriculum materials that contain realistic tasks or activities and normative practices. These include teacher perception of the usefulness of reform, teacher communication of expectations, teacher knowledge of the domain, the use of assessment to guide instruction, and both the long- and short-term impact of that reform on student achievement.

Teacher Perception of the Usefulness of Reform

The initial assessment methods used by teachers at the sites were clearly grounded in their past practices. Despite the efforts of research staff to promote alternative assessment practices, we found that teachers faced difficult challenges in adopting tasks and questioning techniques that reflected reform assessment. In all of the studies, the first step in assessment reform involved motivating teachers to see the need for change, usually by recognizing the quality of work their students were capable of doing (generally as a consequence of changing their instructional practices). All of these teachers were surprised by the work their students were "suddenly" able to do. Burrill (quoted in de Lange, Burrill, Romberg, & van Reeuwijk, 1993) wrote, "The students were excited about ideas—they were thinking and interpreting problems that were real and not contrived. No one said, 'When will I ever need this?'" (p. 158).

Similarly, van den Huevel-Panhuizen (1995) found that when teachers administered open assessment tasks, as one teacher noted, "some of my quieter students displayed a greater understanding than I had given them credit for—some [even] displayed a sense of humor." This same teacher went on to make the following comment about the assessment for the unit: "I feel it offers more than most objective sorts of tests. It allows students to explain their thinking—a very valuable piece of information" (p. 71). Romberg and Shafer (1997) also noted that "in every instance the first thing that struck teachers in analyzing the students' work was that offering rich problems to students results in getting rich answers. This means that simple marking becomes a thing of the past and that giving students credits for 'the' correct answer becomes a hard job" (p. 12).

Overall, as teachers saw the usefulness of reform assessment and its impact on student learning, they moved toward reform material and techniques, noting that they needed and valued the professional development we provided.

Teacher Communication of Expectations

Teachers must realize that normative classroom practices form the basis for the way tasks are used for learning and that they govern the nature of the arguments that students and teachers use to justify mathematical conjectures and conclusions. These norms can be made manifest through overt expectations or through more subtle messages carried, for example, in the questions a teacher asks, which can demonstrate a teacher's genuine interest in understanding students' interpretation of a problem. In particular, teachers and students need to agree on what behaviors and products are to be used for assessment of progress in a domain, and how those behaviors and products are to be judged.

For example, to help students acquire and refine their core mathematics and science concepts, teachers will sometimes require students to articulate their thinking and reasoning, often through the use of written open-ended problems. Accomplishing this, however, does not merely mean adding "explain your answer" or "show your work" to assessment problems found in conventional textbooks (which commonly focus on the solution). To elicit an appropriate response from students, true open-ended problems must be qualitatively distinct from the type of assessment tasks used in conventional mathematics curricula. Otherwise, students will misinterpret the nature of the expected response (Wijers, 2000). Similarly, results from case studies suggest teachers' expectations of student explanations are rarely met. Romberg and Shafer (1997) argued, "It's important for teachers to make their expectations explicit when they ask students to provide explanations of their reasoning or to show their work. When teachers don't provide clear expectations for justification and explanation, students often don't know whether their response is complete or how they should validate their answers" (p. 17).

Teacher Knowledge of the Domain

When teaching reform curricula, teachers sometimes find themselves having to teach a mathematical topic new to them or to the middle school curriculum. At times, they find they do not understand how this topic fits in a broader domain perspective as, for example, an introduction of the tangent ratio in the geometry unit via glide ratios in activities involving similar triangles. When teachers in our studies found themselves in such situations, they were unsure how to proceed, even if they were familiar with the mathematics involved. Most had never taught these concepts to students.

In addition, most units in reform-based curricula involve ideas from several strands and emphasize the interconnectedness of those ideas, an element of instruction very different from teachers' past practices, and quite challenging. This emphasis was also reflected in student strategies and solutions, which proved quite complex to score. Gail Burrill (quoted in de Lange et al., 1993) wrote:

> [We] had to listen to students, examine their work, and try to learn what they were thinking as they solved a problem. What was valid in that process, and what needed to be altered? As teachers, we had to learn how to probe for understanding and explanations; to listen to what students had to say, and figure out when they said what they did. Student work became more important than ever—numerical answers as such had little value. Communication became an integral part of the classroom dynamics. Before the project, we directed our attention to what part of an algorithm students had misused. Now, we had to try to follow a variety of uncharted and often unexpected directions in student work.... Our role was shifting from that of one who directs the thought processes of the students to one who reacts and guides their reasoning; it was not easy to resist telling students what to do or showing them how, but instead to ask them leading questions. (p. 156)

The teachers in these studies varied in their knowledge of mathematics. Many were elementary certified and had little formal course work beyond the traditional high school algebra and geometry courses taken many years ago. Those with undergraduate degrees in mathematics had completed those degrees many years before. None had ever really done mathematics in the manner used in these materials, and they themselves recognized that their lack of background knowledge in the mathematics in the units was a serious problem; in particular, when developing connections within a context:

> The teachers had personally studied mathematics as isolated areas of content with reasonable success, but when it came to integrating mathematics to solve problems within the context of a "realistic" unit, they had difficulty. This led to situations where their mathematical understanding was challenged. (Clarke, 1995, p. 157)

The lack of teacher knowledge in the domains taught is not a problem easily solved. As Lampert (1988) noted, "How can a teacher who lacks a network of big ideas and the relationship among those ideas and between ideas, facts, and procedures develop these things?" (pp. 163–164). We would answer that in our collaborations with teachers, teachers can and have developed instructional practices that addressed this problem, but that they needed and requested support in these areas.

Use of Assessment Information to Guide Instruction

In several studies we found that, over time, teachers developed a more comprehensive view of assessment and began regularly to use a wider range of assessment strategies (e.g., Her & Webb, 2004). The increased attention given to student learning (via assessment) facilitated a greater orientation to the study of relations among content, instruction, and the evolution of student understanding as students progressed from informal to formal reasoning in mathematical domains. In fact, Graue and Smith (1996) contended that the study of classroom assessment can serve as a basis for reorienting teacher practice so that it is flexible and more sensitive to students' understanding of mathematics and science.

The data also gave us insight into the problems and practices around formative classroom assessment. For example, Feijs and de Lange (2004) found that subtle differences in visual presentation of a diagram or use of language in a reform task sometimes caused significant differences in student response. Sometimes these differences determined whether students engaged in the task or problem. In several studies, following implementation of reform curricula, informal ways of assessment (e.g., checking homework, facilitating class discussion, observing student work in progress) had a significant impact on teachers' instructional decisions (e.g., Shafer, 2004; van Reeuwijk & Wijers, 2004). Teachers began to use classroom discourse as a critical construct to negotiate instruction and to assess student understanding (Webb, in press). Teachers often stated that they knew intuitively when they needed information on their students' performance in order to make decisions on further instruction. However, these "professional intu-itions" were still rarely discussed among colleagues unless purposefully brought up in professional development meetings.

Impact on Student Achievement

When teachers conduct formative assessments and give appropriate feed-back, students can learn with understanding. The evidence for this claim comes from three sources: internally developed assessments, quizzes and end-of-unit tests, and externally developed assessments.

Internally Developed Assessments. Classroom teachers at most sites re-ported examples of exceptional student achievement on internally devel-oped assessments. Patterns of teachers' assessment practice evident across these classrooms included teacher attention to student thinking, explicit expectations for all students, and ongoing opportunities to provide feed-back. However, the examples of exceptional performance occurred, as Romberg and Shafer (1997) reported, on specific tasks, on which teachers were surprised that students could do so well.

Quizzes and End-of-Unit Tests. On quizzes and end-of-unit tests, Wijers (2004) reported that when teachers in classes gave a common end-of-unit test with reform tasks, the results on specific items were encourag-ing. Wijers noted, for example:

> Most students developed a broad understanding of the concept of area: broader than often results from teaching that is more traditional, where students tend to identify area with "length times width." This clearly was not the case here. In Problem 1, for instance, the area of irregular shapes was not a problem for the majority of students. If students had known area only as *length × width*, they could not have successfully done Prob-lem 1. The fact that students also were able to construct shapes with a given area (Problem 6) also points to this broad understanding of the concept. (p. 120)

Externally Developed Assessments. The only source of summary data comes from external summary assessments that were administered in sev-eral of the Center's projects. (Information about the impact of a project's ef-forts on elementary school student performance is reported in chaps. 2 & 5, this volume; for detail on summary achievement in the Monona and Verona projects involved in the RAP studies, see chaps. 7 & 10, this volume.) At the Providence RAP site in 1997–1998, the Providence School District began using the New Standards Reference Exam (University of Pittsburgh & National Center on Education and the Economy, 1997), a test designed to show student achievement in mathematics in three areas: skills, concepts, and problem solving. The student population at the school in the RAP study was predominantly minority, with more than 85% of the students eligible for government-funded lunch programs. In all three areas of mathematics achievement on the exam, a greater percentage of students in classrooms of project teachers "achieved the standard" or "achieved the standard with honors" than in matched, nonparticipating classes. With regard to problem solving, the difference between participating ($N = 82$) and nonparticipating ($N = 62$) students that achieved the standard was statistically significant ($p < .05$).

Summary. Across the RAP studies, there was ample evidence that student performance was enhanced in classrooms in which teachers were struggling to implement a reform curriculum. We are unable, however, to separate the effects of changes in instruction from changes in assessment practices, nor should we, because instructionally embedded assessments were part of the program's design.

LONG-TERM DIRECTIONS

The standards-based reform approach to instruction assumes that teachers will use evidence from several sources to inform instruction, but to do this effectively, teachers require support in modifying their practices, learning to develop and score open tasks, and enhancing their knowledge of the domains they teach.

As noted earlier, we are convinced that teachers can learn to use alternative assessment practices as a consequence of appropriate professional development and that, in turn, a professional development program focused on teachers' classroom assessment practices can be used to support changes in teacher practice leading toward student understanding and higher student achievement. But when teachers challenge traditional assessment and work to change their practices and beliefs, they take a risk. Basic to supporting that risk-taking activity are the following:

- A school district and teachers committed to a standards-based reform of its mathematics and science programs, including the commitment of the mathematics and science teachers to building a professional community in relation to changing their formative assessment practices, the willingness of the administration to support teachers as they change their assessment practices, and the willingness of all involved to value the efforts and experimentation of the teachers.

- The availability of a standards-based set of curriculum materials.
- The availability of a long-term, eventually self-sustaining, collaborative professional development program that initially focuses on assessment design, interpretation of student work, and instructionally embedded assessment in relationship to important mathematical goals.
- The importance (understood by all participants) of documenting over time the professional development activities, the changes in formative assessment practices, the impact of changes in assessment practices on changes in other instructional practices, and the impact of overall changes in practices on student performance in mathematics and science.
- Time. Changing practice, learning new techniques, building open collaborations within and across professional communities, learning to focus on what students can do rather than simply determining their rank in the class—all require time to effect, but all these changes can have a significant impact on student achievement.

In the long-term, however, we are aware that any success in reform assessment (and in instruction) still rests heavily on teachers, who take professional and sometimes personal risks in changing their own practices, opening up what they do in the classroom to the view of their colleagues and administrators, and working to enhance not only student "achievement" but also to deepen student conceptual understanding, all the while working to build a foundation for future learning.

REFERENCES

Bachor, D. G., & Anderson, J. O. (1994). Elementary teachers' assessment practices as observed in the Province of British Columbia. *Assessment in Education, 1*(1), 63–93.

Clarke, B. (1995). *Expecting the unexpected: Critical incidents in the mathematics classroom.* Unpublished doctoral dissertation, University of Wisconsin–Madison.

Dekker, T. (1997). *Enquiry project: Assessing realistic mathematics education.* Unpublished manuscript, Freudenthal Institute, Utrecht, Netherlands.

Dekker, T., & Querelle, N. (2002). *Great assessment problems.* Unpublished manuscript, Freudenthal Institute, Utrecht, Netherlands.

de Lange, J. (1987). *Mathematics, insight and meaning: Teaching, learning, and testing of mathematics for the life and social sciences.* Utrecht, Netherlands: Research Group for Mathematical Education and Educational Computer Center (OW & OC), University of Utrecht.

de Lange, J. (1995). Assessment: No change without problems. In T. Romberg (Ed.), *Reform in school mathematics and authentic achievement* (pp. 87–172). Albany: State University of New York Press.

de Lange, J. (1999). *Framework for classroom assessment in mathematics*. Unpublished manuscript. Madison, WI: National Center for Improving Student Learning and Achievement in Mathematics and Science.

de Lange, J., Burrill, G., Romberg, T., & van Reeuwijk, M. (1993). *Learning and testing mathematics in context*. Pleasantville, NY: Wings for Learning.

Evans, F. (1976). What research says about grading. In S. B. Simon & J. A. Bellanca (Eds.), *Degrading the grading myths: A primer to alternatives to grades and marks* (pp. 30–50). Washington, DC: Association for Supervision and Curriculum Development.

Feijs, E., & de Lange, J. (2004). The design of open–open assessment tasks. In T. A. Romberg (Ed.), *Standards-based mathematics assessment in middle school: Rethinking classroom practice* (pp. 122–136). New York: Teachers College Press.

Fleming, M., & Chambers, B. (1983). Teacher-made tests: Windows on the classroom. In W. Hathaway (Ed.), *Testing in the schools: New directions for testing and measurement*. San Francisco: Jossey-Bass.

Frary, R. B., Cross, L. H., & Weber, L. J. (1993). Testing and grading practices and opinions of secondary teachers of academic subjects: Implications for instruction in measurement. *Educational Measurement: Issues and Practices, 12*(3), 23–30.

Fredrickson, A., & Ford, M. (2004). Collaborative partnership: Staff development that works. In T. A. Romberg (Ed.), *Standards-based mathematics assessment in middle school: Rethinking classroom practice* (pp. 188–199). New York: Teachers College Press.

Gerberich, J., Greene, H., & Jorgenson, A. (1953). *Measurement and evaluation in the modern school*. New York: McKay.

Graue, M., & Smith, S. (1996). Shaping assessment through instructional innovation. *Journal of Mathematical Behavior, 25*, 113–136.

Gravemeijer, K., Pligge, M., & Clarke, B. (1998). Reallotment. In National Center for Research in Mathematical Sciences Education & Freudenthal Institute (Eds.), *Mathematics in context*. Chicago: Encyclopædia Britannica.

Haertel, E. (1986, April). *Choosing and using classroom tests: Teachers' perspectives on assessment*. Paper presented at the annual meeting of the American Educational Research Association, San Francisco, CA.

Her, T., & Webb, D. (2004). Retracing a path to assessing for understanding. In T. A. Romberg (Ed.), *Standards-based mathematics assessment in middle school: Rethinking classroom practice* (pp. 200–220). New York: Teachers College Press.

Kilpatrick, J., Swafford, J., & Findell, B. (Eds.). (2001). *Adding it up: Helping children learn mathematics*. Washington, DC: National Academy Press.

Lampert, M. (1988). The teacher's role in reinventing the meaning of mathematical knowing in the classroom. In M. J. Behr, C. B. Lacampagne, & M. M. Wheeler (Eds.), *Proceedings of the 10th conference of the North American chapter of the International Group for the Psychology of Mathematics Education* (pp. 433–480). DeKalb: Northern Illinois University.

National Center for Research in Mathematical Sciences Education & Freudenthal Institute. (Eds.). (1997–1998). *Mathematics in context*. Chicago: Encyclopædia Britannica.

National Council of Teachers of Mathematics. (1995). *Assessment standards for school mathematics*. Reston, VA: Author.

Natriello, G. (1992). Marking systems. In M. C. Alkin (Ed.), *Encyclopedia of educational research* (6th ed., Vol. 3., pp. 772–776). New York: Macmillan.

Romberg, T., & Shafer, M. (1997). *Assessment in classrooms that promote understanding of mathematics.* Unpublished manuscript. Madison, WI: National Center for Improving Student Learning and Achievement in Mathematics and Science.

Senk, S., Beckmann, C., & Thompson, D. (1997). Assessment and grading in high school mathematics classrooms. *Journal for Research in Mathematics Education, 28*(2), 187–215.

Shafer, M. C. (1996). *Assessment of student growth in a mathematical domain over time.* Unpublished dissertation, University of Wisconsin–Madison.

Shafer, M. C. (2004). Expanding classroom practices. In T. A. Romberg (Ed.), *Standards-based mathematics assessment in middle school: Rethinking classroom practice* (pp. 45–59). New York: Teachers College Press.

Smith, M. E. (2000). *Classroom assessment and evaluation: A case study of practices in transition.* Unpublished dissertation, University of Wisconsin–Madison.

Steen, L. (Ed.). (1990). *On the shoulders of giants: New approaches to numeracy.* Washington, DC: National Academy Press.

Stiggins, R., & Conklin, N. (1992). *In teachers' hands: Investigating the practices of classroom assessment.* New York: State University of New York Press.

Stiggins, R., Conklin, N., & Bridgeford, N. (1986). Classroom assessment: A key to effective education. *Journal of Educational Measurement: Issues and Practice, 5*(2), 5–17.

Stiggins, R., Frisbie, D., & Griswold, P. (1989). Inside high school grading practices: Building a research agenda. *Educational Measurement: Issues and Practice, 8*(2) 5–14.

Taylor, C. (1994). Assessment for measurement or standards: The peril and promise of large-scale assessment reform. *American Educational Research Journal, 31*(2), 231–262.

University of Pittsburgh & National Center on Education and the Economy. (1997). *New standards reference examination.* San Antonio, TX: Harcourt Brace Educational Measurement.

van den Heuvel-Panhuizen, M. (1995, April). *Developing assessment problems on percentage.* Paper presented at the annual meeting of the American Educational Research Association, San Francisco, CA.

van den Heuvel-Panhuizen, M. (1996). *Assessment and realistic mathematics education.* Unpublished doctoral dissertation, University of Utrecht, Netherlands.

van Reeuwijk, M., & Wijers, M. (2004). Investigations as thought-revealing assessment problems. In T. A. Romberg (Ed.), *Standards-based mathematics assessment in middle school: Rethinking classroom practice* (pp. 137–151). New York: Teachers College Press.

Verhage, H., & de Lange, J. (1997, April). Mathematics education and assessment. *Pythagoras, 42,* 14–20.

Webb, D. C. (2004). Enriching assessment opportunities through classroom discourse. In T. A. Romberg (Ed.), *Standards-based mathematics assessment in middle school: Rethinking classroom practice* (pp. 169–187). New York: Teachers College Press.

Westbury, I. (1980, January). Change and stability in the curriculum: An overview of the questions. In *Comparative studies of mathematics curricula: Change and stability 1960–1980* (pp. 12–36). Proceedings of a conference jointly organized by the Institute for the Didactics of Mathematics (IDM) and the International Mathematics Committee of the Second International Mathematics Study of the International Association for the Evaluation of Educational Achievement (IEA). Bielefeld, Germany: Institut für Didaktik der Mathematik der Universität Bielefeld.

Wijers, M. (2000, April). *Explanations why? The role of explanations in assessment problems*. Paper presented at the annual meeting of the American Educational Research Association, New Orleans, LA.

Wijers, M. (2004). Analysis of an end-of-unit test. In T. A. Romberg (Ed.), *Standards-based mathematics assessment in middle school: Rethinking classroom practice* (pp. 100–121). New York: Teachers College Press.

Wilson, L. D. (1993). *Assessment in a secondary mathematics classroom*. Unpublished doctoral dissertation, University of Wisconsin–Madison.

Chapter 13

Capacity for Change: Organizational Support for Teaching for Understanding

Adam Gamoran
University of Wisconsin–Madison

The organizational context in which most teachers work is designed to support predictable routines (Rowan, 1990), yet teaching for understanding can take off in unexpected directions, placing substantial new demands on teachers and on the contexts in which they work. To support teaching for understanding, schools will need to increase their capacity for change, and we can expect confusion and conflict about aims and approaches to be a major issue in schools undergoing organizational change. In our work examining school and district contexts, therefore, we take a broad view of resources, identify teacher professional development as the primary engine of change, and view school organization as a dynamic system, both influencing and influenced by the programs implemented in it (Gamoran, Secada, & Marrett, 2000). We propose that schools can best support teaching for understanding by *responding* to teacher learning: allocating substantial time for professional development, supporting teacher autonomy in instructional content and pedagogical methods, and allocating resources in ways responsive to teachers' efforts. The goal of the work reported here was to identify and understand the supports and barriers in the school and district contexts of teachers attempting to teach for understanding.

Our findings rest mainly on qualitative analyses of interviews and observations from six professional development group sites: four in Wisconsin (suburban elementary and middle school in the same district; suburban high school; urban high school), one in Massachusetts (urban K–8 elementary); and one in Tennessee (urban middle school). At each site, we observed professional development (over 100 sessions total), interviewed teachers (most twice, for a total of 150 interviews) and administrators (several twice, for a total of 35 interviews), and surveyed over 150 teachers (66–86% response rate; survey findings were used to supplement qualitative evidence). Our findings are discussed in this chapter around the key issues of access to resources, leadership of organizational change, and sustainability. (For greater detail, see Gamoran et al., 2003.)

ACCESS TO RESOURCES

Taken as a whole, our findings indicate that the relation between professional development and resources is a two-way process. Teaching for understanding occurs when teachers develop new habits of practice through sustained, cohesive professional development and have access to resources and structures that allow their insights to flourish. Professional development that emphasizes teaching for understanding can, in effect, transform material resources into new human and social resources, in that teachers gain new knowledge and skills and strengthen their relationships with other educators. These newly created resources, along with additional resources from outside venues, make it possible for professional development to affect teachers' classroom practice. At each site we studied, we found that allocation of time, use of curricular and professional development materials, and availability of internal and external expertise all had an impact on this process.

Resources That Support Professional Development

Time. From the perspective of teachers, time was the most important material resource, specifically, time for teachers to work together on issues of student thinking in mathematics and science. But even when financial resources were available to provide extra time for teachers to work together, practical limitations made finding time to meet difficult. One option, meeting during the day, meant that substitutes covered classes. In at least three of our sites, however, teachers were frustrated with the quality of their substitutes and found it difficult to miss class time on a regular basis. The other option, meeting after school, became a regular solution at four of the six sites, but competing demands on teachers' after-school time, including other school and district projects and obligations, made this solution problematic.

In the Wisconsin urban high school, teachers met during their common planning periods during the school day, but that solution meant that project time competed with other planning issues for teachers' attention. By contrast, professional development time at the other sites was paid time *added on* to the normal work schedules, and scheduled planning time was often used to share with their colleagues what they were doing in professional development. This solution appeared most common in the Wisconsin suburban sites, probably because of the extensive time teachers had to meet. The elementary school teachers had two half-days per month for professional development or other teacher meetings, plus one hour per week of planning time. The middle school teachers had two periods each day, one to meet with their team and another for individual planning. At the high school, science department members met daily. Regular teacher meetings and school sponsored activities thus became occasions for diffusion of ideas and materials drawn from professional development, creating a sustained base of support that extended beyond the participants.

Curricular and Professional Development Materials.   The importance of material resources in supporting teachers' efforts to apply what they learn cannot be overstated, and finding funds that enable that flow of resources can be vital to reform. Teachers at the Wisconsin elementary site, for example, were enthusiastic about TERC curricular materials, and funds from the district and a university research grant permitted them to purchase these books and supplies. At the Massachusetts elementary site, when lack of materials from the district and schools impaired teachers' ability to carry new ideas back to their classrooms, funds from the teacher–researcher professional development collaborative made implementation possible. In the Massachusetts district, however, teachers of language-minority students generally had more difficulty obtaining materials than did regular classroom teachers, in part because innovative bilingual curricular materials are simply less available. As teachers in *bilingual education*, however, they also lacked access to special district resources set aside for teaching for understanding in science. These challenges were compounded by the usual problems of obtaining basic supplies in a large urban district, resulting in teacher perceptions of isolation. This district placed great weight on teacher entrepreneurship, an approach gaining currency in education reform circles. Under this approach, teachers had to rely on their own initiative to obtain resources, exploiting their connections and relationships as well as developing and articulating innovative ideas. Those who tended to be less successful entrepreneurs, however, such as novice teachers or those outside established networks, often failed to obtain the resources they needed.

Overall, we found prepackaged curriculum materials less essential for supporting professional development than the new knowledge provided by outside experts. At only one of the six sites, the Wisconsin suburban middle school, was a prepared curriculum central to the professional development program. In other cases, developing curricular approaches and materials constituted much of the work of the professional development collaboratives, and this helped teachers to match their curricular interests to their students' emerging thinking. At the Wisconsin suburban high school and elementary school, for example, teachers produced tangible curricula subsequently implemented in school classrooms. The elementary teachers also wrote up classroom projects, some of which were compiled as material resources and used by other teachers in later years. For teacher initiatives with less access to outside experts, unlike the six we studied, prepackaged curricula might prove more important as resources.

Internal and External Expertise. Precisely because teachers' expertise and knowledge about student thinking was limited, outside expertise was essential to stimulating teachers' initial investigations and learning. University researchers served as key resources at all six sites. The Wisconsin elementary site is an interesting case in point. In the first year we interviewed teacher participants, 30% listed at least one member of the research team as someone they would go to with a question. In the second year, the percentage more than doubled, to 63%. In the third year, after the researchers had pulled back from their leadership role, one third of the teachers still said they called on a member of the research team when a question arose.

Although in the successful elementary and middle school cases in Wisconsin and Massachusetts, district officials played key roles in promoting the teacher–researcher professional development groups, in the long run the substantive expertise of the teachers themselves (created or developed through professional development) proved even more important than administrative support, as this suburban high school teacher noted:

> People that I work with—that resource is invaluable. I don't know what I would do in another situation where I didn't have the experience of the other teachers to draw on for ideas and things of that nature. That resource is the one without which I would be floundering.

The growth in teacher learning was clearly evident at these sites, and teachers commented on their learning in interviews. This finding would be entirely unremarkable were it not for earlier research showing that teachers often do *not* learn from professional development and consider it a waste of time (Fullan, 2001). More important than the learning of individual teachers, however, was the increase in the collective level of knowledge and expertise among teachers within a school or district. Though we cannot demonstrate this increase in human resources with the present data, it seems probable that the more expert teachers teaching in the school, the more likely other teachers, even those not part of a professional development group, benefited from the collective wisdom. Teachers examined student work collaboratively, discussed what students were thinking and how their thinking changed in response to classroom experiences, and together worked on activities that would enhance their students' understanding. This collaborative process fostered group cohesiveness and over time built professional community. Teachers came to rely on each other, as noted in this exchange after teachers in the Wisconsin middle school site realized the researchers did not have the time to work with all the teachers who desired their support:

Teacher 1:   That's the main thing that you miss [from] student teaching. Bouncing ideas off of other people. It helps me to think about things.

Teacher 2:   Last year, [the lead researcher] shared a lot with me. One of the biggest advantages of being in the project is having an extra adult and being able to share observations on what you did.

Teacher 3:   Last year, there seemed to be a better balance. With a bigger group, it's harder.

Researcher: Last year, the number of teachers was equal to the number of researchers. With everyone doing teaching experiments, there are more teachers than researchers.

Teacher 3:   Would it be possible to work together *without* researchers? [Teacher 4] and I observing each other would in some ways be more valuable.

Summary

Changes in teaching are dependent not only on teacher learning, but also on the material, human, and social resources that make it possible for teachers to act on their own new knowledge and commitments. Any resource stretched thin or unavailable, whether time, materials, or outside expertise, has an impact on professional development as well as on teaching, learning, and student understanding.

Clearly, availability of time for teachers to meet together is the most indispensable material resource, but human and social resources in the form of teacher expertise and leadership, a strong preexisting colleague group (or the creation of community through professional development), supportive school and district administrators, or outside linkages are important to the creation and maintenance of ongoing professional development. But we note that adequate resources and teacher commitment cannot bring about teaching for understanding in isolation. At the Tennessee site, for example, resources were available from both the school and the district, and teachers and researchers were interested, but sharp conflicts between district goals and the purpose of teacher–researcher collaboration forestalled collaboration, and the professional development group program did not go beyond the planning stage.

We also note that human and social resources, whether generated or allocated, can be decimated through administrative changes and turnover within the school district. The professional development programs we observed in the Wisconsin suburban elementary and middle schools had great success in generating resources, but these gains were tempered by the departure of key teachers from the elementary site and by administrative decisions to shift teachers across schools at both sites, in part due to a burgeoning student population (and the subsequent creation of new schools). These moves fragmented social networks established (or enhanced) through years of teacher–researcher collaboration.

SCHOOL LEADERSHIP FOR CHANGE

Leadership in a context of teaching for understanding must nurture new mechanisms for responding to the uncertainties of teaching. Commonly, teachers avoid uncertainty by adopting scripts or predictable patterns of practice that allow them to carry on with teaching as if it were a routine activity with predictable consequences (Jackson, 1968). When teaching emphasis is on student thinking, however, teachers are forced to examine their assumptions and to respond to students' ideas in their daily work. In Rowan's (1990) view, this situation calls for a more organic approach to management, in which leadership responds rather than constrains the technical core (teaching and learning) through the allocation of resources and, as Spillane, Halverson, and Diamond (2001) noted, through supporting shared, or distributed, leadership.

"Managing" Change

Allocating Time.   Teachers' views of administrative barriers to success with teaching for understanding focused, in part, on time—either time that they needed to devote to other tasks or time denied them when they wanted to be working on understanding student thinking. At one site, for example, project teachers wanted to use a designated school-wide professional development day to participate in a workshop with a nationally known expert on science education. The only school contribution necessary was the release of the teachers from the school-wide activity, yet the principal initially refused. Eventually, a compromise permitted the teachers two hours to meet with the national expert. Many principals we spoke with recognized this issue as a dilemma and noted that they felt compelled to use professional development time for school-wide issues. They regarded allocation of time to issues of concern to the design collaboratives as a sacrifice.

Facilitating Teacher Learning. From the perspective of many teachers, simply staying out of the way was one of the most important contributions that principals could make to their efforts to focus on student thinking in mathematics and science. Teachers viewed the absence of barriers as a form of support even when they claimed their principal had little idea of what they were doing in their professional development groups. We found, however, that supportive principals did more than just avoid posing barriers. As one urban principal in Massachusetts explained, "I look at myself as [taking] a facilitator, cheerleader role, not an expert role." His comments were echoed by a suburban principal in Wisconsin who said, "I see my role as a facilitator, someone who creates the environment where good teaching can take place and where decisions can be made in the best interests of kids." These and other supportive principals allocated resources in *response* to new directions that teachers identified. Rather than pushing a particular approach, they served as linkages, helping teachers with common approaches find one another. As one senior district administrator explained, "Our model of governance is not just shared decision making, it's shared leadership.... What is the role of principal then? The role changes to coordinator–facilitator, lead by model, not by directive ... coach, partner, creating learning opportunities for people [so they can create] circles of excellence."

A principal in the same district described her experience with this type of leadership:

> You go into administration with the idea that you'll be able to control and have an effect on what happens with that school. And the biggest lesson for me in all of this is the best way to control it is ... you stand on the sidelines, and you say, "That's great! Good job! Would you like to try this next? Here's something else you can do. Did you know so and so was doing this?" As opposed to saying, "This is how it's going to be done. You've got until Friday to turn this in." That way doesn't work. If you want people to behave as professionals, you have to treat them as professionals. And that means they make the majority of the decisions, and they listen to each other, work things out.

Fostering School Vision. Principals did play a role in fostering school visions, but the visions they expressed were not specific to mathematics or science. We found more administrative involvement in setting content-specific visions at the district level, primarily from specialists in curriculum and instruction. At one site, a district-wide vision of mathematics or science was reflected in a specific curriculum that teachers were expected to follow. Whether or not this approach can support teaching for understanding depends on the content and direction of the curriculum. In the Massachusetts urban site, a district director of science promoted a curriculum that was geared toward teaching for understanding. Consequently, teachers who joined the professional development collaborative, which emphasized scientific reasoning among students of diverse backgrounds, could participate and remain fully consistent with the district's direction. By contrast, the Tennessee site mandated a core curriculum that was not at all compatible with the teacher–researcher collaborative's emphasis on in-depth understanding of a limited number of powerful mathematical ideas, and teachers were required to follow closely a prescribed sequence of instruction and testing. As one teacher noted:

> Curriculum changes, it just won't happen this year. We've got too many people looking at us, you know, coming from the district, they're going to come out, check our records, what we're doing to see that we are implementing city curriculum, and it's not just, "Oh, yeah, we've got it," you know, that we are [really] doing it.... Everybody's got to follow the little guidelines. (Interview with Tennessee urban middle school teacher)

In this case, not only was teaching for understanding unsupported, but teachers felt the district mandate did not leave room for the explorations offered by the professional development groups.

Supporting Teacher Autonomy. At the Wisconsin suburban district, an entirely different model was evident at both the suburban elementary and the suburban middle school sites. In this case, administrators emphasized the need for teachers to pursue their individual passions, developing "circles of excellence" around commitments to particular practices. Although this approach highlighted differences among teachers and created problems of inconsistent approaches across the district and even within the same schools, it also provided motivation, opportunity, and, when complemented by material and human resources from inside and outside the district, strong support for a teacher-led approach that emphasized teaching for understanding. District administrators claimed they personally favored "constructivist" teaching approaches, but their broader commitment was to allow teachers to identify their own sense of what constituted excellent teaching, and to pursue that vision with vigor.

We noted that strengthening teacher autonomy can make it difficult to establish a coherent direction for curriculum and instruction within the district and school. The Wisconsin suburban elementary site, which placed the greatest emphasis on autonomy in developing "circles of excellence," faced striking differences in the approaches that different groups of teachers wished to take. In some schools, a student might encounter a teacher with a strong constructivist approach to mathematics in one year and a teacher who placed greater emphasis on drill and practice in the next. But as one elementary principal explained, "We have a lot of control over the hiring policy within the building, so our interviews are structured to find people who have a similar philosophy."

Extensive hiring opportunities are not often a viable option, however, and promoting a common approach across the district might create more coherence. Indeed, the Massachusetts urban district's promotion of a district-wide approach to teaching science came in response to earlier lack of coherence. Our analysis of the trade-off is supported by results of the teacher surveys. The teachers at the Massachusetts and Wisconsin sites we surveyed reported extremely high levels of classroom autonomy, approaching the maximum score for "a great deal of influence" on these items. Teachers in the Wisconsin district also exhibited extraordinary influence over school policies; in contrast, the Massachusetts teachers reported less influence and were more comparable to the national average. This pattern was also seen in Wisconsin responses about shared leadership. We also note that, compared to the national averages, both the Massachusetts and the Wisconsin teachers less often regarded their principals as setting clear expectations, knowing what type of school they wanted, and communicating that to the staff. In our data, the opening of opportunities for teacher autonomy was associated with less vision-setting by the principal. (See Gamoran et al., 2003, for more details of the survey results.)

Fostering Distributed Leadership

The Role of the Researchers. Teachers generally came to the professional development collaboratives with an interest in teaching for understanding, but *visions* of what teaching for understanding meant emerged through the collaboration of teachers and researchers, particularly as they viewed and discussed teachers' classroom practices and their students' work. At the meetings, the researchers played a variety of roles, from guiding teachers through activities to remaining more on the sidelines, but they invariably appeared ready to jump in when the teachers had questions or needed focus. In interviews, teachers expressed appreciation for this approach to leadership, commending the researchers for validating teachers' ideas and actions, and appreciating their openness in asking and answering questions.

Teacher Leadership.   At all of the active collaboratives, teachers frequently identified colleagues as leaders because of their experience, knowledge, and skills. Teacher-leaders themselves preferred to lead by example rather than by the exercise of authority, even when they held official leadership positions. The creation of such positions was essential, however, so that management tasks such as creating a schedule, disbursing funds, maintaining contacts with district administrators, and setting agendas for meetings could be accomplished, especially in the cases in which the researchers were trying to turn "ownership" of the professional development groups over to the teachers. These new positions, which embodied both management and leadership, required an investment of resources from the research group and ultimately from the school district, because teachers expected to be paid for taking on administrative tasks.

Sustaining and diffusing teaching for understanding was an explicit goal at the Wisconsin suburban sites, but not at any of the urban sites. Not surprisingly, therefore, there was more emphasis on teacher leadership at the suburban sites. Still, important aspects of distributed leadership involving teachers were evident at two of the urban sites. In the Massachusetts teacher–researcher collaborative, a few teachers were consistently identified as leaders by their peers. In this case, a combination of strong district support and relatively weak school-level involvement left room for teachers to act as entrepreneurs, acquiring resources through internal and external grants to support their efforts beyond the resources provided. Moreover, leadership opportunities through the professional development group might have been particularly important for the teachers of bilingual classes, who (as noted earlier) generally saw themselves as isolated within their schools. At the Wisconsin urban high school site, teachers also used the professional development collaborative to set and pursue an agenda important to them. Although the project was short lived, it did not end due to lack of teacher leadership. Rather, the school's division into families resulted in fragmented departmental bonds, and the social divisions within the school between bilingual and monolingual teachers prevented cross-school ties that would have made a sustained program possible. Both of these cases contrasted with the Tennessee site, the only case in which we found no evidence of distributed leadership in support of teaching for understanding. Teachers there were held strictly accountable for a mandated curriculum, and principals were constantly aware of central office supervision.

All three of the suburban Wisconsin sites showed evidence of stable teacher leadership. Interview responses to questions about leadership were corroborated by observations of professional development, in which teachers played leading roles even early on, showcasing their work with students. At the high school, the boundary between teacher and researcher was blurred; one of the teachers held a part-time position in the research institute, and one of the researchers took a part-time job in the school. In the middle school, established teacher-leaders used the professional development collaborative to suit their purpose of creating heterogeneous mathematics classes in Grades 6–8. In the elementary group in the same district, leadership was widely distributed among teachers from four different schools. We observed at least 10 different teachers displaying their students' work to the group, and even more were named as leaders by their colleagues in the interviews.

Summary

Organic management is not the bureaucratic leadership of directives and supervision. Leaders in districts that supported teaching for understanding played facilitative roles, linking together teachers who could pursue common goals and making it possible for teachers to work with outside experts. Supportive leaders also fostered distribution of leadership among a variety of participants, particularly teachers, who then played active roles in leading their own learning.

SUSTAINING TEACHING FOR UNDERSTANDING: LOOKING AT A SUCCESSFUL COMMUNITY

Sustainability in teaching for understanding refers to teachers not only utilizing what they have learned in professional development in their teaching, but also continuing to grow and develop in their understanding of student thinking and in how this understanding informs their practice. Franke, Carpenter, Levi, and Fennema (2001) referred to this second type of sustainability as "generativity," in that the focus on teaching for understanding generates a continued growth in knowledge and understanding, with teachers not only maintaining new practices over time, but continually modifying and adapting practices in response to new learning and their ongoing reflection on student thinking.

In this section, we briefly examine our evidence from the Wisconsin suburban elementary site, the only one for which we have clear evidence of sustainability. Two researchers and 10 teachers initiated this group, which grew to include as many as 34 elementary teachers at its height, with over half of the district's regular classroom elementary teachers participating at

some point. In the year after the researchers had reduced their involvement, 27 teachers still participated regularly, and teachers from this professional development community continued to meet as a group or in small
groups for at least two years beyond that time.

Resources

The teacher meetings were funded by the district—at some points this meant
paying teachers to attend meetings after school; at other points it meant paying for substitutes so teachers could attend during school. The district also
provided funding for a teacher to coordinate the professional development
collaborative.

Teachers in this group clearly exhibited the characteristics of a professional community: Shared values, collective focus on student learning, collaboration, reflective dialogue, and the open sharing of their teaching
practices were all evident during observations of professional development.
The level of integration within this group was very high. In their interviews,
teachers frequently noted that they greatly valued their opportunities for
collaboration, but they also expressed three concerns about maintaining a
well-integrated professional development group:

1. Identifying leadership within the group to manage the logistical tasks
   necessary for ongoing meetings and collaboration.
2. Maintaining an acceptable group size. Some argued that the group had
   become too large to satisfy the needs of all members; others argued
   that, if necessary, they could split into smaller groups arranged by interest, grade level, or subject matter. (During the final year of our observations, when the teacher-led group had grown to 27 participants, the
   group did split into smaller groups.)
3. Diminishment of the group through turnover. Although some staff
   left, the group survived turnover and transitions, including teachers
   being involuntarily transferred from one school to another within the
   district. (Our evidence suggests that although such changes present
   real challenges, they need not spell the end of community.)

Building Linkages

This professional development collaborative had strong linkages with the district and school administration. Two principals each spent a year attending the
collaborative's workshops, and a key district administrator was instrumental
both in securing district funds ($15,000 annually) to support the program and
in convincing teachers to take on group leadership after the researchers reduced their involvement. There was also a high potential for linkages among
teachers within and across schools. By the time our observations were drawing

to a close, a few teachers expressed a desire to expand to include as many teachers as would like (or could be convinced) to participate. These teachers were motivated by a desire to address problems of coherence in pedagogical approach within one elementary school. They also wanted to reduce the isolation of two participants now in another elementary school. Other teachers responded skeptically to these proposals, wondering if the program would become diluted as it expanded or would become unmanageably large.

Supporting Autonomy and Distributed Leadership

Interviews revealed a consistent philosophy among school and district administrators of developing "circles of excellence," a notion that places a premium on teacher autonomy. At the same time, the district used choice—charter school options and different programs within the elementary schools—to defuse potential objections to one direction or another that teachers might collectively take. We note that in this collaborative and the surrounding district, state content standards tended to energize rather than constrain teachers and that teachers referred to the standards when deciding what "big ideas" or powerful mathematics and science content they wanted as the focus of their efforts to teach for understanding. Despite the issues of philosophy, the increasing diversity in the student population, the teacher reshuffling in part due to the continual growth of the district, and the mix of pedagogical approaches, the quality and continuity of the professional development group were maintained.

Summary

Analysis of this case shows that a professional development community focused on teaching for understanding can be sustained beyond the initial investment of a teacher–researcher collaborative. Beyond continuing the necessary flow of material resources, sustainability of reform approaches rests on the strength of distributed leadership, in which teachers play a major role, and on strong linkages with district and school administrators. Even the most dedicated teachers generally cannot, by themselves, sustain the resources and the commitment necessary for teaching for understanding: They need the support and advice of colleagues, and they need continuing access to new developments in their fields.

CONCLUSIONS

Our analysis of the evidence collected at these sites suggests that rich, collaborative professional development is the most productive investment a school district can make. High-quality professional development can generate new resources in the form of human and social capital for teachers, and, unlike investment in equipment, textbooks, and training, this investment does not depreciate.

Supporting this investment, however, means that commitments will need to be aligned, most likely in one of three ways:

- Districts can mandate a district-wide approach to teaching in mathematics and science. (Our data suggest that this approach can either support or inhibit teaching for understanding, depending on its content.)
- Districts and schools can encourage, but not mandate, an approach (e.g., with a recommended curriculum) and work to achieve coherence by selecting staff comfortable with the intended focus. (Most schools, however, are not in position to withstand substantial staff turnover and simultaneously select staff who buy into a specific vision.)
- Districts can encourage both teacher autonomy and school-wide distributed leadership, provide resources that enable professional development in accordance with that autonomy, and package it all with an extensive choice system for parents. (In light of our data, this seems to be the most workable solution.)

This last response will mean that districts need to establish a compelling vision in support of teaching for understanding, provide teachers with the autonomy to enhance that vision or develop their own, and work to support distributed leadership. Newmann and Associates (1996), in their study of highly restructured schools, noted that of the two most successful cases of promoting authentic pedagogy, in one (similar to the Massachusetts site), educators subscribed to a common vision that supported inquiry and depth, and in the other (similar to the Wisconsin elementary site), teachers were expected to develop their own "circles of excellence." As noted, we saw both these instantiations at the sites we studied. At these sites, teachers also took on leadership roles, primarily by developing expertise. In a supportive context, their new knowledge, skills, and relationships with colleagues naturally found an outlet in mentoring and leading by example. But because bureaucratic tasks still needed to be fulfilled, even in a context of distributed leadership, teachers whose expertise brought them into informal leadership found themselves called on to carry out formal leadership tasks as well. Our findings suggest that both informal leadership in teacher inquiry and formal leadership in managing logistics are needed to sustain a professional development group and to disseminate reform practices beyond the original participants.

We emphasize that any district response will need to be supported by infusions of resources from both inside and outside the district in order to complement and sustain the human and social resources developed in professional development. On their own, administrators (and teachers) seldom have the expertise to bring about a deep focus on student understanding. The Wisconsin elementary case suggests, however, that if material resources can be maintained by the district, the human and social resources can become self-generating, even after the outsiders end their involvement.

AUTHOR NOTE

This chapter is based on work reported in detail in *Transforming Teaching in Math and Science: How Schools and Districts Can Support Change*, by A. Gamoran, C. W. Anderson, P. A. Quiroz, W. G. Secada, T. Williams, & S. Ashmann, 2003, New York: Teachers College Press. Copyright 2003 by Teachers College Press.

Research was conducted under the auspices of the National Center for Improving Student Learning and Achievement in Mathematics and Science, supported by funds from the U.S. Department of Education, Office of Educational Research and Improvement (Grant No. R305A60007), as administered by the Wisconsin Center for Education Research, University of Wisconsin–Madison. Findings and conclusions are those of the authors and do not necessarily reflect the views of the supporting agencies.

REFERENCES

Franke, M. L., Carpenter, T. P., Levi, L., & Fennema, E. (2001). Capturing teachers' generative change: A follow-up study of teachers' professional development in mathematics. *American Educational Research Journal, 38,* 653–689.

Fullan, M. (2001). *The new meaning of educational change* (2nd ed.). New York: Teachers College Press.

Gamoran, A., Anderson, C. W., Quiroz, P. A., Secada, W. G., Williams, T., & Ashmann, S. (2003). *Transforming teaching in math and science: How schools and districts can support change.* New York: Teachers College Press.

Gamoran, A., Secada, W. G., & Marrett, C. B. (2000). The organizational context of teaching and learning: Changing theoretical perspectives. In M. T. Hallinan (Ed.), *Handbook of research in the sociology of education* (pp. 37–63). New York: Kluwer Academic & Plenum Press.

Jackson, P. W. (1968). *Life in classrooms.* New York: Holt, Rinehart & Winston.

Newmann, F. M., & Associates. (1996). *Authentic achievement: Restructuring schools for intellectual quality.* San Francisco: Jossey-Bass.

Rowan, B. (1990). Commitment and control: Alternative strategies for the organizational design of schools. *Review of Research in Education, 16,* 353–389.

Spillane, J. P., Halverson, R., & Diamond, J. B. (2001). Investigating school leadership practice: A distributed perspective. *Educational Researcher, 30*(3), 23–28.

Postscript

Thomas A. Romberg
Thomas P. Carpenter
University of Wisconsin–Madison

Learning with understanding is at the heart of mathematics and science reform. The fundamental issues in reforming mathematics and science teaching and learning revolve around what mathematics and science are most important for students to learn, how this content can be taught most effectively to all students, and what is required to implement and support this kind of instruction in schools.

The chapters in this book illustrate how norms for modeling and argumentation emerge in classrooms, and how these forms of reasoning contribute to the development of mathematical and scientific understanding. Instruction that focuses on developing understanding and that engages students in the mathematical and scientific practices of modeling and argumentation requires teachers to make ambitious and complex changes. Professional development and support for teachers engaged in reforming their practices is critical, and that support must go beyond simply providing them with new curricular materials. Schools need new conceptions of teacher professional development that move beyond traditional training–coaching models. As Little (1993) pointed out:

> The dominant training model of teachers' professional development—a model focused primarily on expanding an individual repertoire of well-defined and skillful classroom practice—is not adequate to the ambitious visions of teaching and schooling embedded in present reform initiatives. (p. 129)

Showing teachers how to implement effective practices is not enough. Teachers need to understand why they are being asked to do what they are shown. We need to understand what it means for a teacher to engage in on-going learning as well as how professional development and the development of professional community can contribute to that end. The conceptual framework for the research on professional development described in this book is grounded in our conceptual analysis of teacher learning with understanding that is, or can be, ongoing.

In the same way we think about students learning with understanding, we conceive of teacher change in terms of teachers acquiring knowledge that can support inquiry, deepen student problem solving, and provide a basis for acquisition of new knowledge and continued growth. Professional development must enable and support teacher inquiry (into subject matter, student learning, and teaching practice), so that teachers can adapt their practices in ways appropriate to the demands of subject matter and their students' learning, and judge student performances appropriately. The research in this book provides support for treating teachers as professionals who have the capacity to transform their teaching practices in a generative fashion, over time.

To engage in classroom-based inquiry, teachers need a conceptual framework for understanding what students are saying, how that reflects what the students know, and what implications what students know has for instruction. This classroom-based inquiry must be connected to disciplinary knowledge oriented toward "big ideas" in mathematics and science. Focusing on the structure of the disciplines helps teachers to reconceptualize their notions of science and mathematics, to use the notion of learning corridors to guide their instructional plans, to deepen their analysis of students' learning, and to validly assess students' understanding. For most teachers, this approach is revolutionary in its acknowledgment that mathematics and science must be rethought and cannot simply be assumed to be well captured in traditional curricular scope and sequences.

The research in this book shows teachers engaged in ongoing inquiry to deepen their own mathematical and scientific understanding, their understanding of development of their students' mathematical and scientific conceptions, and the instructional and assessment practices that foster that development. But our research also shows that teachers' inquiry does not survive well in isolation. The development of professional communities is critical to sustaining and generating teacher change, for many of the same reasons that mathematicians and scientists conduct their work within larger communities of inquiry. The creation and maintenance of teaching communities is critical for reform because such communities provide a climate for engaging in inquiry, sharing knowledge of student thinking, sharing norms for what counts as effective instruction and student achievement, and building social supports for managing uncertainty. Although strategies for creating communities to sustain long-term teacher professional development vary widely, these strategies all involve substantial restructuring of schooling in order to enhance collaboration between teachers and administrators. This restructuring is aimed at assisting teachers in developing the resources necessary to conduct practical inquiry and in sharing the results with a larger community.

Inquiry and sharing the fruits of that inquiry are at the heart of the profession. To foster high-quality instructional programs in mathematics and science, teachers need to share learning goals, take collective responsibility to reach them, jointly address the challenges that arise, and share in developing methods to respond to those challenges. Only collectively can teachers influence school policy. Research describing cases of school improvement point to professional communities of teachers as necessary for effecting organizational change to support teaching and learning for understanding, but if instructional practices are to be self-sustaining, communities of learning need to be expanded beyond classroom teachers. The classroom instruction of teachers has the most direct effect on student learning, but the school, district, and community can have a profound effect on what teachers are able to implement in their classrooms and the support that they receive for continuing to participate in meaningful, ongoing professional development. Reform of teaching and learning must be accompanied by serious efforts to help teachers engage in dialogue and build partnership with the wider community of administrators, parents, and citizens.

REFERENCE

Little, J. (1993). Teachers' professional development in a climate of educational reform. *Educational Evaluation and Policy Analysis, 15*(2), 129–151.

Author Index

Subject Index

Milton Keynes UK
Ingram Content Group UK Ltd.
UKHW022100141024
449569UK00031B/1718

9 780805 846959